THE GREAT WINE BLIGHT

OTHER BOOKS BY GEORGE ORDISH

Untaken Harvest (1952)
Wine Growing in England (1935)
Garden Pests (1956)
The Living House (1959)
The Last of the Incas (with Edward Hyams, 1962)
Man, Crops and Pests in Central America (1964)
Pigeons and People (with Pearl Binder, 1967)
Biological Methods in Crop Pest Control (1967)

Translations

Hong Kong to Barcelona in the Junk 'Rubia', by José María Tey (1962)
The Mask of Medusa, by Roger Callois (1964)
The History of the Incas, by Alfred Métraux (1969)
Animal Societies, by Remy Chauvin (1968)
The World of Ants, by Remy Chauvin (1970)

THE GREAT WINE BLIGHT

George Ordish

WITH EIGHT PAGES OF PLATES,
ELEVEN ILLUSTRATIONS IN TEXT
AND A MAP

Charles Scribner's Sons · New York

A – 7.72(I)

Printed in Great Britain
Library of Congress Catalog Card Number 72-537
SBN 684-10438-5

Contents

Illustrations

PLATES BETWEEN PAGES 98 AND 99

IN TEXT

Preface

This is the story of one of the triumphs of pest control that started more than a hundred years ago and saved wine for the world. I am much indebted to many people and books for help in writing it. I would like to thank them all, in particular:

Ascot, Berks, Miss J. A. Ellis, J. Kennedy, R. F. Skenty; *Ay, Champagne*, G. Montgolfier; *Bonn*, Herren Köble, Steinlein; *Bordeaux*, R. Pijassou, D. Schvester, P. Sichel, G. Schÿler; *Boston (Mass.)*, Miss E. Kastonotis; *Cambridge (Mass.)*, Mrs C. E. Rubin, S. J. Schneider; *Chapingo, Mexico*, N. Sánchez Durón; *Château Loudenne*, M. Bamford; *Cognac*, N. Burnet, M. Philiou; *East Grinstead*, the late André Simon; *Harpenden, Herts*, M. Cohen, H. W. Jansen, H. Stroyan; *Haslemere, Surrey*, Mrs Wenban-Smith; *Jerez de la Frontera*, M. Gonzalez; *Kew, Surrey*, R. G. C. Desmond, Sir G. Taylor; *Leagrave, Beds*, A. R. Toms; *London*, Mrs M. A. Cousins, M. Durodier, Mrs Peta Fordham, M. Friend, S. F. Hallgarten, V. Hargreaves, H. Hopf, E. Hyams, J. Jeffs, Miss O. Moran, J. Nimmesgern, Miss E. J. Orton, R. Pope, Mrs C. Richmond, V. L. Seyd, J. Stevens, J. A. Smith, H. E. Thrupp, the Misses J. and W. Weiner; *Manchester*, E. R. Povall; *Marseilles*, M. Rubert; *Montpellier*, MM. Bernon, Clave, Lafert, Labri; *Nashville, Tennessee*, R. P. Warwick; *New York*, Martin Dockery, Mrs Meliora Dockery; *Oxford*, R. Cobb; *Patchogue, Long Island*, Miss C. V. Riley; *Paris*, G. Baissette, R. Colas, R. Dion, J. B. Fergusson, A. Gassier, A. J. Hays, Mlle A. Lazansky, Mme L. Mabille, F. Tomlin; *St Albans, Herts*, R. Gordon, G. Pazzi-

Axworthy; *Sandgate, Kent,* Mrs O'Brien, Mrs L. René-Martin; *Sudbury, Suffolk,* G. E. Fussell; *Trencrom, Cornwall,* Mrs C. Shipton; *Washington D.C.,* Mrs H. E. Edwards.

I am also much indebted to help from the Librarians of the Bordeaux Archives and Municipal Library, the Boston (Mass.) Public Library, the Cambridge (Mass.) Harvard Library, the British Museum, Foreign Office, Patent Office and Tropical Products Institute Libraries, London, the Royal Botanic Gardens' Library, Kew, the School of Agriculture Library, Montpellier and the New York Public Library.

I apologize in advance for any omissions in this and finally would like to thank my wife for much help and useful comment.

George Ordish

Conversion Factors

The majority of the figures used in this book are in metric units and French francs.

The conversion factors are:

1 kilometre	=	0·621 miles		
1 hectare (ha.)	=	2·471 acres		
1 hectolitre (hl.)	=	22·00 British (Imperial) gallons		
		26·418 U.S.A. gallons		
1 metric tonne	=	2,204·62 lb.	=	1·102 short tons
			=	0·984 long tons avoirdupois

1890 Exchange rates

	per £	per U.S. $
Belgium	25·50 francs	5·24 francs
France	25· 21 francs	5·18 francs
Germany	20·67 marks	4·25 marks
Italy	25·86 lire	5·31 lire
Netherlands	12.175 guilders	2·50 guilders
Portugal	52·25 escudos	10·74 escudos
Spain	45·00 pesetas	9·24 pesetas
Switzerland	25·525 francs	5·24 francs
U.S.A.	4.867 dollars	—

The small superior numbers in the text refer to the bibliography. Notes (pp. 194–204) are indicated by a reference in brackets thus: (Note 12).

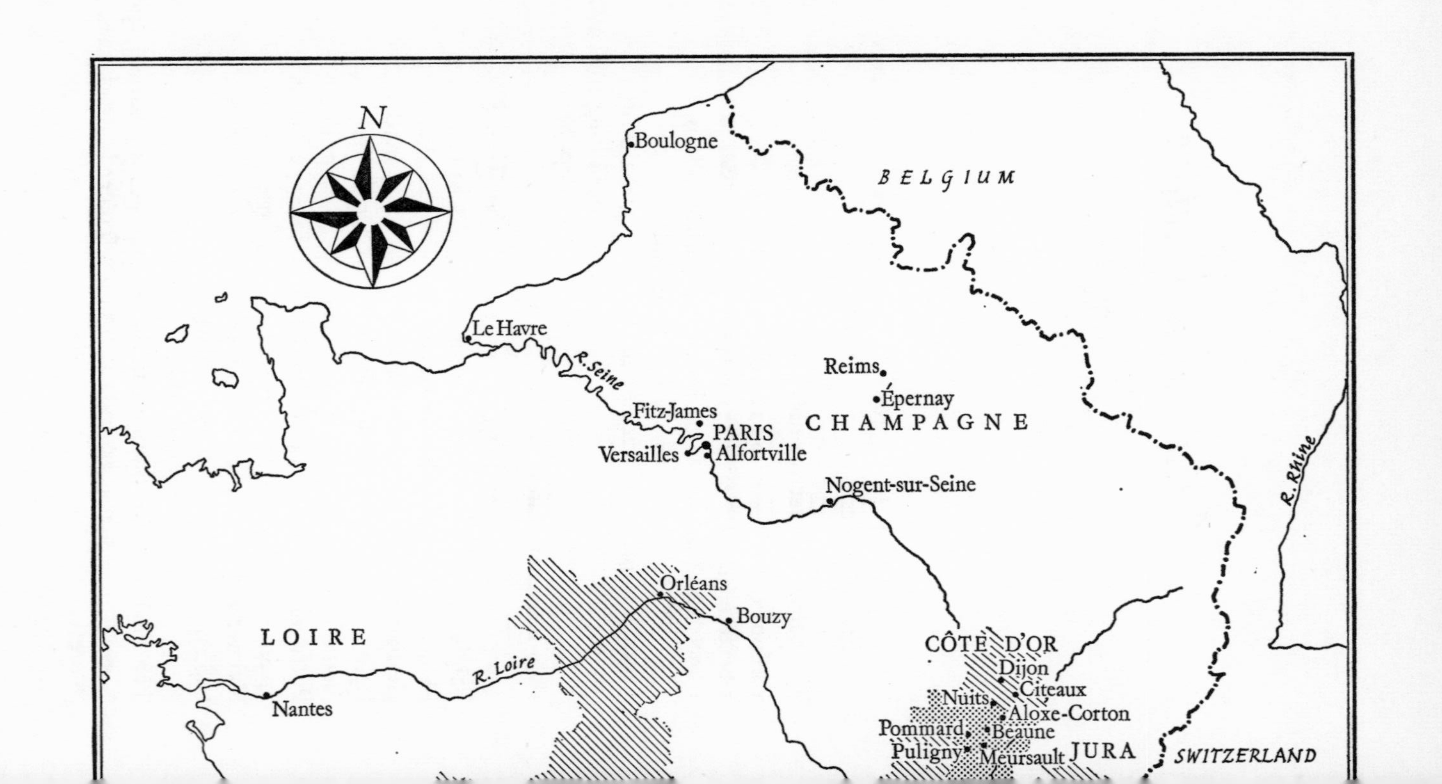
N
Boulogne
BELGIUM
Le Havre
R. Seine
Reims
Épernay
Fitz-James
PARIS
CHAMPAGNE
Versailles
Alfortville
Nogent-sur-Seine
R. Rhine
Orléans
Bouzy
LOIRE
R. Loire
Nantes
CÔTE D'OR
Dijon
Citeaux
Nuits
Aloxe-Corton
Pommard
Beaune
Puligny
Meursault
JURA
SWITZERLAND

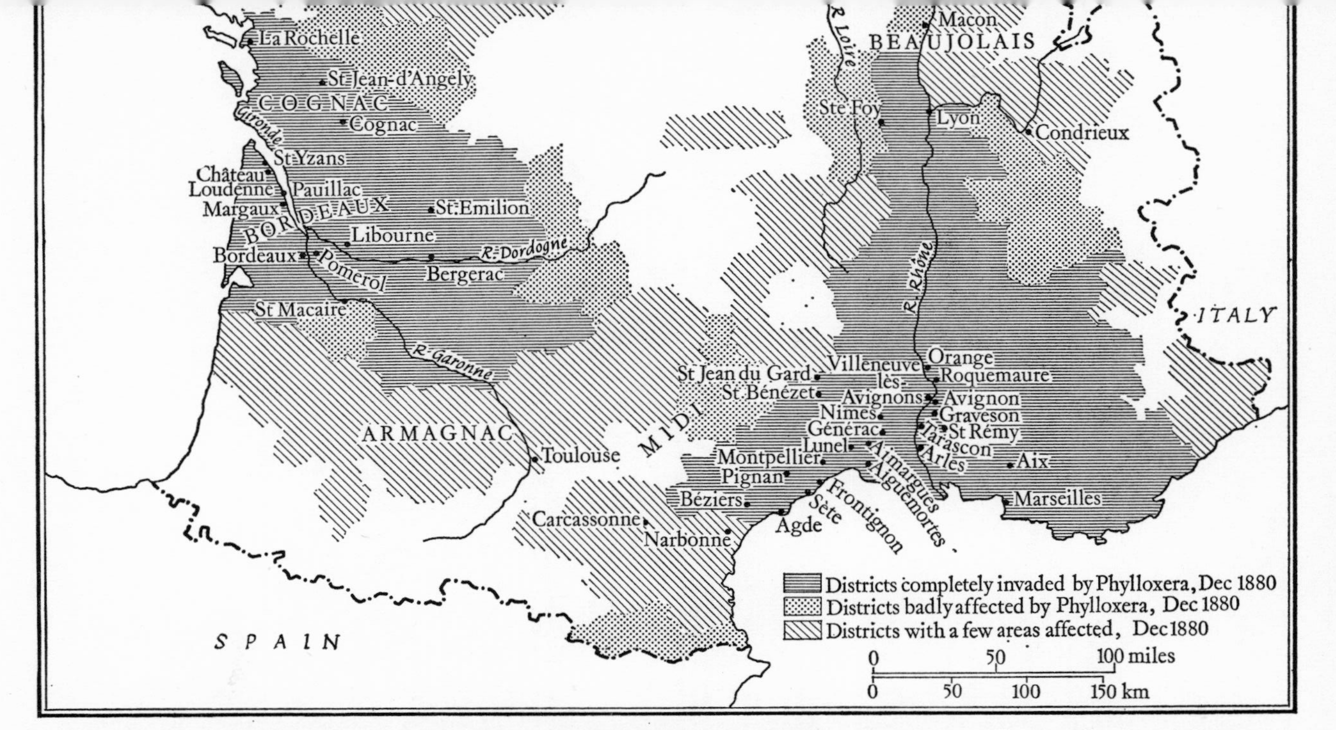

Phylloxera in France. Based on the map issued by The Phylloxera Commission in 1880. Eventually the whole country was invaded.

CHAPTER ONE
Due Warning

In 1863 there occurred in England an event which did not in the least disturb the brilliant world surrounding the Emperor Louis Napoleon in Paris. He was unlikely to have heard of it and certainly would not have worried about it if he had, yet it was to cause more economic damage to France and Europe than Louis's mad attack on the Prussians seven years later. Even in England it passed unnoticed.

This event was first manifested by the sending of a sample of insects from a greenhouse in Hammersmith to Professor Westwood, the famous botanist and entomologist at Oxford University. The insect was the phylloxera. Coming from America, it slowly and surely killed any European vine to which it gained access. It would thus have wiped out the wine grape throughout Europe and eventually the whole world. But in 1863 even Professor Westwood did not regard the newly discovered insect as anything more than an entomological curiosity. Six years later he published an account of it: 'In the month of June, 1863, I received from Hammersmith a Vine leaf covered with minute gall-like excrescences, "each containing," in the words of my correspondent, "a multitude of eggs, and some perfect Acari, which seem to spring from them, and sometimes a curiously corrugated Coccus."'[186] This book is the story of what the insect did, what it might have done and how the threat was overcome.

In 1863 Hammersmith was a fashionable suburb to the west of

London, on the north bank of the Thames. It contained many splendid houses belonging to rich business men. In 1863 every city merchant's house was large, had a garden and the essential conservatory. These had become increasingly fashionable ever since the success of Sir Joseph Paxton's enormous Crystal Palace for the Great Exhibition of twelve years earlier. If we may believe the *Punch* picture of that age it was the only place where, amid the exotic plants and shrubs, Edwin could steal a kiss and pop the question to Angelina who, after a due delay, would refer him to papa. The place thus played an important part in the propagation of the species and the consolidation of the family and its fortunes. Edwin would never get into the conservatory, certainly not alone with Angelina, were he not considered suitable as a husband.

Thus the greenhouse had to be properly stocked and it tended to be filled with foreign plants. To achieve this, nurserymen ransacked the world both for new curiosities and for cheap stocks of old favourites.

Many now common plants were first sent over by the intrepid plant hunters during this period. The 'Ward case' enabled the more delicate ones to travel—together with their pests, naturally. Nathaniel Bagshaw Ward (1791–1868) found that certain plants would thrive for long periods under a bell-jar or in a closed glass container of one sort or another, a fact much exploited by fashionable shops today. The closed glass carboy full of strange fern-like plants growing on and on, apparently forever, is but the Victorian 'Ward Case'. In this closed system there seems to be enough CO_2 production to provide considerable leaf growth and even flowers at times. The Ward case protected plants carried on deck from sea water, wind and torrential rain and enabled them to get light; it allowed such men as Robert Fortune to bring the forsythia, the primula and many kinds of chrysanthemum from Japan. Sanders of St Albans scoured the world for orchids and made a fortune from them. A wide range of plants became available, running from the difficult orchids to the Virginia creeper which, once planted, grew itself. Some of the strange vines from America also were imported. They were decorative, grew quickly, did not need much skilled attention and even at times produced some rather coarse fruit.

The business was profitable and ranged from £33,000 (1859)

to £81,000 per year, declared customs value;[176] there was no import duty.

All these plants, including the novelties, were imported with no regard whatever to the possibility of introducing insect pests or diseases. Man was then so egocentric that the idea of such insignificant forms of life being a threat had not really penetrated to the minds of politicians or the general public, though there had been plenty of examples of the damage such rivals could cause: the potato blight (*Phytophthora infestans*), for instance, starved a million people in Ireland; the chestnut blight, introduced on plants from Europe, wiped out its host tree in America. Even the vines themselves had suffered from the introduction in the 1840s of the powdery mildew. The age had many distinguished biologists (Darwin, Huxley, M. J. Berkeley, John and William Curtis, etc.), yet little notice was taken of biologists in the everyday business of life, possibly because the world believed in the 'practical man', *laissez-faire* economics and the power of manufacturing and business. Biology was mostly conducted by dilettante amateurs who did not disturb the basic belief that everything had been created by God for man's benefit, though it was hard sometimes to see how. Such questions as 'What is the *good* of the bedbug?', if such an indelicate matter could have been broached, would have been answered by 'We shall never know in this life, sir.'

Another agent in the spread of pests during this period was the collector of insects, mostly amateur entomologists with the collecting mania for small detail seen today chiefly in the keen philatelist, for although people still collect insects, and show great interest in them, devotees do not now display the fanaticism of the nineteenth century. John Curtis, the distinguished English entomologist, earned his living for many years by supplying 'gentlemen's cabinets' with specimens until he was able, with great relief, to turn to the more practical side of entomology.[50] The craze was equally great in France; for instance, Monsieur J. Desbrochers des Loges, of Tours, in the 1890s put out a printed price list of Coleoptera alone with some 9,000 named species in it. Many insects were especially bred for the market. In Great Britain certain suppliers (not John Curtis; he had too much sense of responsibility to do such a thing) were actually breeding the Colorado beetle (*Leptinotarsa decemlineata*) as a valuable curiosity for their customers and it is quite remarkable that this serious potato pest did not become established in Europe during the mid nineteenth century. The first British legislation against crop pests was especially designed to prevent entomologists breeding the Colorado beetle. M. Desbrochers des Loges did not have this particular insect on his 1895 list, but he had a large number of other pest species there, such as the American cotton-boll weevil ('Anthonomus de l'Amérique'), the turnip gall weevil (*Ceutorynchus napi*) and the turnip flea beetle (*Phyllotreta brassicae*), but who the big spender was who was going to pay fifteen centimes for such a common, widespread pest it is difficult to say.

M. Desbrochers des Loges accepted payment in French or foreign postage stamps and also entered into elaborate exchange arrangements by means of his journal *Le Frelon*.*

Darwin in 1859 had startled the world by publishing *The Origin of Species by Means of Natural Selection*, and by 1863 garbled versions, all equally inaccurate, were still being debated in college common rooms and London pubs, the 'descent of man from monkey', much emphasized by the press, having caught the public's attention, though the 1860 meeting of the British Association for the Advancement of Science at Oxford did much to explain the new ideas. In fact the meeting can be considered a turning point. Oxford also, amid its 'dreaming spires', was greatly disturbed. On the one hand, by becoming converted to the Roman Catholic faith Newman had alarmed that world almost as much as had Darwin on the other, with natural selection.

William Irvine [99] puts this dilemma very neatly. 'Newman's conversion . . . had opened an abyss of conservatism on one side; now Mr Darwin's patient and laborious heresy had opened an abyss of liberalism on the other.' At this 1860 British Association meeting Darwin pleaded illness and Professor Huxley was persuaded to be there to answer the expected attack on Darwinism from a practised speaker, Bishop Wilberforce (known to the irreverent as 'Soapy Sam'). The bishop spoke with skill and forceful rhetoric, carrying the meeting with him. Huxley noted that Wilberforce had but little idea of what the theory was about and he (Huxley) kept very quiet. Flushed with success, the bishop

* 'The Hornet', a strange name for a publication devoted entirely to beetles.

turned towards Huxley and in ending '... begged to know, was it through his grandfather or his grandmother that he claimed his descent from a monkey?' The applause was tremendous. Huxley rose and explained the theory calmly and scientifically, touching on the bishop's ignorance and, still grave, concluded that he would not be ashamed to be descended from a monkey, but that he would be 'ashamed to be connected with a man who used great gifts to obscure the truth'. The sensation and applause were again enormous and Darwinism was launched as a force that influenced all future biology. To mark the occasion at least one lady present fainted.

Other new thoughts and developments were astir, and the successful London World Exhibition of 1862 celebrated the view that the introduction of free trade and the Manchester school of economic thought had been the right road to follow.

Probably the most important technical development of the 1860s was the increase in the efficiency and reliability of the steamship, which undoubtedly led to the spread of pests around the world as the pace of international commerce quickened. The early steamships used vast quantities of fuel and were unsuitable for long journeys. Whereas Captain Sir John Ross,[157] on his amazing voyage to discover the North-West Passage, 1829–33, threw his engine away when he got to Magazine Island, the vessels of the 1860s (for example the *Lancing* and the *Pierre*, two whose times are documented) could steam 360 nautical miles a day, thus regularly crossing the Atlantic in some nine or ten days.

The time taken for the Atlantic voyage before the coming of the steamship explains why the phylloxera did not reach Europe, or at least was not established there, sooner than 1863. The insect is so destructive that a long interval between its presence and its discovery is unlikely, so that its successful establishment cannot have been much before 1863. American vines were introduced to Europe well before this critical date. Loudon [116] lists the introduction of various *Vitis* species from 1629 onwards (*see* Table 1, page 207).

These early vine introductions are likely to have been cuttings, rooted cuttings and seeds. If they had been infected with phylloxera aphids they would have died by the time the long sea voyage had been completed. But the steamship carried the plants far more quickly and the railway reduced the time of the inland voyage from port to garden, so that the pest survived its long journey and, once established, spread rapidly.

At this same time (1863) in France an 'unknown disease' was being talked about. Two slopes of the Bas-Rhône had been attacked and by 1867 it had reached such proportions in Corntat, Crau (Bouches-du-Rhône), parts of the Alps and around Tarascon that general alarm was spreading among the *vignerons*. They had good reason to be worried, for many of them remembered the crisis caused by the 'Oidium', a fungus disease which, coming from America in the 1840s, nearly extinguished viticulture as a commercial enterprise (*see* Chapter Three). But the new disease was very unlike the 'Oidium', for there appeared to be no cause. With the new trouble a vine or two, usually in the centre of a vineyard, would start to sicken, the leaves yellowing at first, the edges then turning red, the leaves finally drying up and dropping. The next year the symptoms would be worse and could be seen spreading to neighbouring vines; the extension growths were weak, the dried-up tips being easily broken in winter. If any fruit set it might ripen if the attack was slight, but would be of poor quality, watery, acid and with no bouquet. The black grapes stayed clear red or pink in colour. Wine made from them was very poor and did not keep. With heavier attacks the bunches of grapes just dried up and usually fell off. The third year the vine was dead, and when dug up the external tissues of the roots were black and rotting and could be pushed off with pressure from a fingernail to expose the fibrous core.

The symptoms reminded people so vividly of 'consumption' in man, the *phtisie*, that killed so many promising young men and women, particularly in the romantic novels of that age, that it was often referred to colloquially as the *étisie*. The leaves even had the unhealthy red flush so often seen on the cheeks of the sufferer from consumption. It was quite natural and logical to think of it as a disease.

The real cause was not identified until 1868, and then it was hardly noticed in Paris amidst the political struggle which had been in progress for some years. Louis Napoleon had been declared emperor in 1852, after his third *coup d'état*, and in spite of the fact that he had solemnly sworn to maintain the republican constitution. Two successive referenda confirmed his position; he dissolved the National Assembly and ruled as dictator. About the time the phylloxera was making its first landing in France the Emperor was turning towards the establishment of a more liberal régime.

The Great International Exhibition was held to manifest the brilliance and self-confidence of the Empire and opened on 1st April 1868. *L'Illustration* [98a] struck a warning note about the current optimism and self-confidence: 'Peace and Prosperity are fine, as is the Temple of Peace at the Exhibition, but every country in Europe is arming. As we open the Temple of Peace Europe is ordering three million guns.' The majority believed in the prospect of peace, prosperity and material success and this optimism was reflected in the interest of all society in the Exhibition and in huge attendances. The machine gallery created enormous interest; in other pavilions exotic plants were displayed and prizes given. Agriculture was represented by an annexe exhibition at Billancourt. Of course no quarantine was imposed on any plant material in the Exhibition and it is surprising that future dangerous pests were not imported. But these were of little moment in the ebullient atmosphere of Paris in 1868.

The Exhibition was a success; the liberal empire was becoming established, along with the phylloxera, and the Iron Chancellor of Germany was nothing like so bad as he had been painted.

CHAPTER TWO
The World of the Vine

Though not essential to the phylloxera story it may be of interest to and complete it for many to know something of the botany of the various kinds of grapevines before going further, but this chapter may easily be omitted by those wanting to get on with the story.

There are many species of vines with a wide range of characteristics. The European grapevine was named by Linnaeus as *Vitis vinifera sativa* L. and is the only vine species giving grapes capable of being made into palatable wine. It is a very ancient plant in the family Vitaceae and has been derived from an ancestor, *Vitis sezannensis*, now found only as a fossil, which flourished in Western Europe in the Lower Eocene epoch. Botanists include in the Vitaceae four genera which bear edible fruits, though, apart from *Vitis*, not very palatable ones: *Ampelopsis*, *Cissus*, *Tetrastigma* and *Vitis*. The Virginia creepers, formerly *Ampelopsis*, are now classified as *Parthenocissus*.

Planchon, as quoted by Foëx and Vialla,[76] split the genus *Vitis* into two sub-genera, the *Muscadiniæ*, with only one species, *Vitis rotundifolia* Michaux (*see* Note 1), and the *Euvites*, the true grapevines containing the European *V. vinifera* and all the other species of similar botanical structure. Planchon's classification is set out in Table 2, page 207.

There are some forty-eight species of grapevines giving fruit of any economic importance, thirty-five in America, twelve in Asia and one in Europe, our *Vitis vinifera*. *Vinifera* seems to have arisen on the shores of the Black Sea. There are hundreds of *vinifera* cultivars and hybrids, with a wide range of differing characteristics. As early as 1952 Sr Lorenzo Badell R.[12] issued a list of varieties which came to some 500 titles, and possibly as many new ones have arisen since then.

Botanically the fruit of the grapevine is an indehiscent, superior berry, the second adjective referring to its position relative to the flower, not its quality. Superior berries, such as the grape and the tomato, grow above the flower and inferior berries, such as the gooseberry and cucumber, grow below it. 'Indehiscent' means that the fruit does not burst or split open to discharge the seed, examples of dehiscent seeds being peas and beans. Mature grapes for wine are found in tightly pressed bunches; obviously they are not and could not be thinned to produce large, separated berries as are the hot-house table grapes. One has to avoid confusion between the French word *grappe* (the tightly pressed bunch of fruit, such as vines and grapefruit produce) and the English word 'grape', the fruit itself. It is curious to notice that the grapefruit is so named because it is produced in bunches or *grappes*; though the French for grapefruit is *pamplemousse*.

The flowers of the vine are many, small, green and inconspicuous. In the wild state the vine is mostly diœcious, that is it has separate male and female flowers, but hermaphrodite flowers are occasionally found, having stamens and ovary in the same flower. Practically all cultivated varieties of vine are hermaphrodite, it being almost essential to cultivation that they be so. This is an interesting instance of early, unconscious human selection and we cannot do better than quote Mr Edward Hyams on this point:

If a population of wild vines, or cuttings from them, had been planted by some prehistoric gardener, he would have noticed that some vines bore no fruit at all (males), some bore much or little, unreliably and uncontrollably (females), and that a third class—the vines with a proportion of hermaphrodite flowers—always bore some fruit and could be regarded as reliable. The gardener, if a sensible man, would have chosen the first for destruction, the third for propagation. But the male vines being grubbed and destroyed, the females would have received less pollen than before, and the crop from them would have fallen, showing up the hermaphrodite vines even more clearly as the most useful class.' [96]

In the perfect, or hermaphrodite, vine flower the calyx is formed of five rudimentary sepals and there is a corolla of five rudimentary petals joined at the base, forming a small hood covering the five stamens and the ovary (*see* Fig. 1). Sometimes the flowers have six

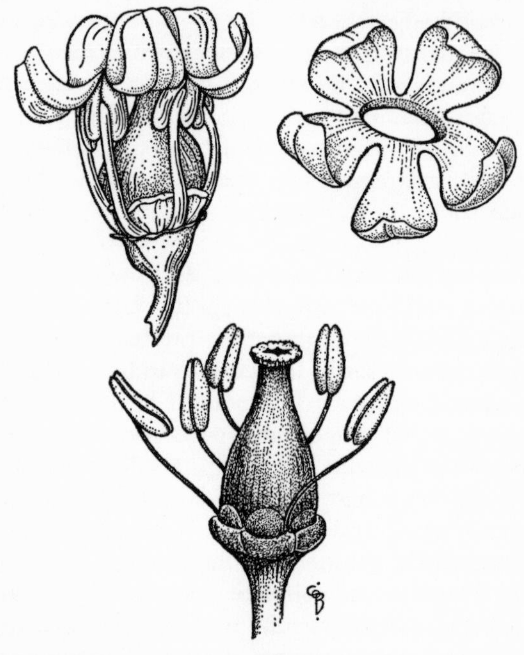

1. *Hermaphrodite vine flower (cultivated vines).*
TOP LEFT: *Flower petals unfolding from the top, an unusual arrangement.* TOP RIGHT: *The solid ring of petals which falls from the flower.* BOTTOM: *Stigma and pollen-bearing anthers of flower. The anthers spring away from the flower when the petal ring falls.*

petals. Between the stamens are a number of little swellings called nectaries, secreting a sugary fluid having a special perfume, particularly noticeable in a vineyard on a warm, calm day. The flower opens from the bottom, the petals curling upwards from their central ring, which is unusual in flowers, and the ring, or hood, of five (or six) joined petals is pushed off by the growth of the stamens and ovary beneath it. In a vineyard during flowering there is quite a rain of these tiny petal rings onto the soil. After the petals have fallen the incipient berries look *smaller* than before the flowers opened, a fact often surprising the newcomer to the vine. The stamens carrying the anthers and pollen are pressed against the ovary and stigma in the closed flower, but when the petal ring is pushed off the stamens spring away from the ovary and the anthers open to discharge pollen, which is thus unlikely to fall on the stigma of its own flower—no incest in this world—this cross pollination being a factor in the production of so many species and varieties.

After pollination the fruit swells rapidly. The berry has two segments each containing two seeds and may be of any colour from greenish white through rose to purple. In most 'black' grapes only the skin is coloured. Red wine is made by crushing black grapes and leaving the must to ferment on the skins, when the alcohol formed extracts the colour from the skins. Much white wine is made from black grapes (most Champagne is, for instance), the fruit being taken straight to the press after picking, when only white juice (known as 'must') runs out. Nevertheless there are some varieties of grapes having coloured flesh; they are called *teinturiers* and the colouring substance is soluble in water, and thus is distinct from that of the skin, which only dissolves in alcohol.

The leaves of the vine tend to be alternate, and opposite the leaf stalk a tendril or a bunch of grapes is usually found; in fact the fruit bunch is a modified tendril and tendrils themselves sometimes bear a few grapes. The fruit is thus borne on the current year's growth. Vines can grow to an immense size and a great age and we quote two records. Joshua's spies brought back from the land of Canaan a bunch of grapes so big that it needed two men to carry it (Numbers xiii, 23); and a staircase of the temple of Diana at Ephesus, one of the seven wonders of the ancient world, was said to have been cut out of a single immense vine stem (Note 2).

Foëx and Vialla [76] by 1886 had summarized the American grapes in a scholarly work, already mentioned in connection with Planchon's classification, from which we extract a few more details. *V. rotundifolia* Michaux, the only species in the Muscadinia, was known in America as the Muscadine, Bullace or Bullet grape. From its leaves it would seem to be the existing species most like the fossil *V. sezannensis*. *Rotundifolia* is a southern species and does not grow much north of the Potomac (38° N.). It is much found wild in woods, its vines climbing over the tallest trees. The best-known cultivar was Scuppernong, with yellowy-bronze coloured fruit; others were Flowers, Thomas and Mish, with violet-coloured fruit, and Tender Pulp, Richmond and Pedee. The cultivars are very vigorous, and Planchon refers to a single plant in North Carolina that covered an area of about two acres. The berries ripen consecutively and drop as they mature.

The Euvites section has four species with large berries and fifteen species with small fruits, the most important species of the former being *V. labrusca* and of the latter *V. æstivalis* and *V. riparia*.

Labrusca is known as the 'Fox Grape' because of its rather repellent 'foxy' taste, and is sometimes called, possibly out of politeness, blackcurrant or strawberry. Nevertheless *labrusca* still has its *aficionados* for eating, juice and wine; its cultivars Concord and Catawba continue to be much grown in the U.S.A. Bush and Meissner, the nineteenth-century American nurserymen [36] [37] [38] divided the *labrusca* varieties into two groups; (i) those from the north, very vigorous with very 'foxy' fruit, the chief variety being Concord (Foëx and Vialla [76] quote sixteen others, mostly unknown today); (ii) those from the south, the chief cultivars being Catawba and Isabelle (Foëx and Vialla give twelve more, and add a third group of fourteen varieties of 'unknown origin', none of them found much today).

Æstivalis, as the name suggests, is known in America as the 'summer grape'. It is not a heavy cropper and there are three types: (i) lobed leaves, bunch-winged and long, containing varieties such as Jacquez and Herbemont; (ii) almost whole leaves, stocky bunches, short and unwinged, containing such varieties as Cunningham and Black July; and (iii) where the veins of the leaves are covered with down on the underside, containing varieties such as Cynthiana and Hermann.

Bush and Meissner [38] divided them into three groups: (i) Southern, containing Black July, Cunningham, Herbemont and Jacquez; (ii) Northern, containing Cynthiana and two others; and (iii) unknown origin, containing five little-known cultivars.

Riparia is found extensively in America, occurring from ninety miles north of Quebec to 28° N., in Kentucky, Illinois, Missouri, Arkansas, Colorado, New Mexico and south Utah. Meissner said he had not seen it in Louisiana or Texas, but the Texans are said to have called it the River Grape and the Sweet-scented Grape, from the perfume given out by the male flowers. The fruit is small, early and somewhat foxy. The main varieties were Taylor, Clinton and Winslow.

Rupestris was another of the American species of *Vitis* found in Missouri, Arkansas, Indian Territory and New Mexico, and was known as the Sand Grape in this first state. It was not grown as a crop. *Berlandieri* was well known in Texas under the name of Sweet Mountain. It was introduced to Europe by M. Douysset, suffering a sea change to Surret Mountain in the process.

Hybrids are now very numerous. Those best known in 1885 were: (i) (mostly *riparia*) Clinton, Black Hambourg, Vialla; (ii) (mostly *labrusca*) York-Madeira, Noah, Othello; (iii) (unknown origin) Brant, Scuppernong, Schuylkill. Foëx and Vialla list many others.

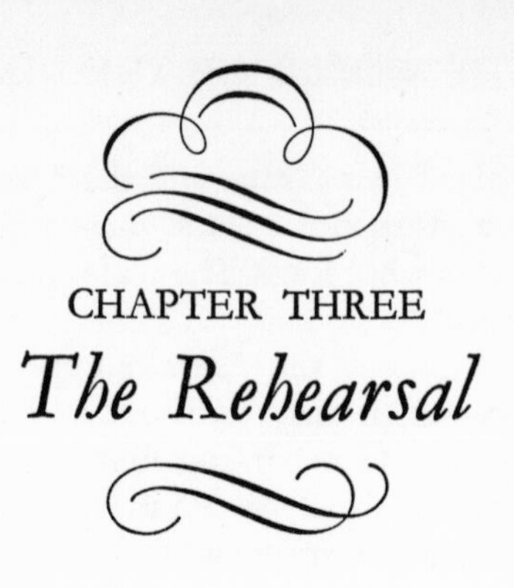

CHAPTER THREE
The Rehearsal

Fate allowed France in particular, and Europe in general, a rehearsal for the phylloxera attack; this was the powdery mildew of the vine. The experience of overcoming this disease was to stand the country in good stead when the far more serious phylloxera came along, because many scientists and administrators who fought the insect had been through a similar situation with the Oidium, as the new disease was first called.

In the 1840s the Reverend M. J. Berkeley was a poor curate at King's Cliffe, Northants. One presumes that his rector was an understanding man, because the young Berkeley spent a great deal of time studying fungi, particularly those that caused diseases in plants; in fact Berkeley became famous as the inventor of the new science of 'vegetable pathology'. He began to get diseased plants from all over the world, which may well have been the cause of spreading a number of pests.

In 1845 a Mr Tucker, gardener to Mr John Slater at Margate, Kent, sent some diseased vine leaves to the fungiferous curate for identification and advice on controlling this trouble in his vine house. Berkeley decided it was a new species and named it *Oidium Tuckerii*, in honour of the gardener, and wrote an account of it which appeared in the *Gardeners' Chronicle and Agricultural Gazette* for 1845.[21] It was wrongly named because Berkeley had seen only the vegetative form. The sexual stages were not found until 1892 (three years after Berkeley's death), when the fungus was found to belong to the genus *Uncinula* Lév. and given the specific name of *necator* Burr.

The common name is very descriptive; it is a white powdery mildew growing on the surface of leaves, stems and fruit. The attacked parts of the plant fail to grow, dry up and fall. Diseased fruit cracks and remains small and acid. The spores blow about in the wind, or are carried away by the rain, and set up new infections, much reducing the vigour of the attacked plant. The ascigerous form, that is the winter resting stage, is seldom found in Europe.[9]

Whence it came to Britain is by no means certain, but it was no doubt brought in on some ornamental plant, vines or Virginia creeper. It occurs on native plants of this nature in both the United States and Japan. An American origin of the fungus is suggested by the fact that American varieties of vines, such as Isabella, Catawba and York-Madeira, show some resistance to the disease. Moreover the ascigerous form is quite common in America. Attempts made by a few botanists of that age to show a European origin were not very convincing. That it was not an adaptation of a powdery mildew of the bindweed, as was thought at one time, was shown by the fact that diseased bindweed was often found climbing up perfectly healthy vines and healthy bindweed found growing on infected vines. Pliny's *Natural History* has much about the troubles of the vine, but his descriptions are too vague for a positive identification, and the same applies to Targioni-Tozzetti, writing in 1766.[171]

In Europe the fungus mostly survives the winter by means of resting cells found in the vegetative mycelium.[9] It was soon being found in France. A. M. Grison, head of the forcing houses at the Palace of Versailles, was particularly troubled by its attack on the royal grapes in 1846, and in 1847 Mr J. Rothschild reported it at Suresnes. In 1848 Grison managed to retain his post when the grapes became republican again and were no less vulnerable for that. He started to look for, and eventually found, a remedy. The powdery mildew was soon attacking vines all over France. In 1851 the southern French vineyards were badly attacked, and the disease had been noted also in Spain, Italy, Greece, Hungary, Asia Minor, Switzerland and Algeria. Alarm was general.

By 1852 the disease was endemic in all the vineyards of Europe and North Africa,[135] but the grapes in M. Grison's greenhouses at Versailles were kept healthy, not so much because they had now become imperial, but because Grison had invented the spray known today as lime-sulphur but then called *Eau Grison*. Spraying with lime-sulphur, made by boiling lime and sulphur together and pouring off the clear yellow liquid, was a practical proposition for greenhouses, where small syringes and 'garden engines' could be used, but it could not be done on a vast scale in huge vineyards. Even so, attempts were made to do it, and the *Revue Horticole* in 1855 suggested that fire engines should be used, with the nozzles suitably adapted with a rose or other device to throw a fine spray.

Grison became something of a hero.[91] At Versailles the vines flourished and the Prefect of the province said M. Grison should be rewarded.

Spraying being quite impractical, the losses suffered by the vine-growers became enormous and panic ensued. M. Henry Marès (who later had much to do with the phylloxera) reported that vineyards at Lunel (near Montpellier), had lost 90 per cent of their crop and even more had been lost at Frontignan. There was a great shortage of wine, which led to a price increase, so that the financial loss to growers was not as severe as it might have been. Moreover the powdery mildew, unlike the phylloxera, did not kill the vine. Nevertheless much misery was caused and people started to leave the wine areas.

Dusting with fine sulphur was possible on a field scale, and this alternative process was put in hand by Monsieur Duchâtre. It was highly successful, and by 1863 there was hardly a vineyard that was not treated. So successful and important was the remedy that a whole 'science of sulphur' arose. The reason for its success was studied by M. Marès, who also found that sulphuring both increased the set of fruit and led to an earlier harvest, a great advantage. For instance, in the Gironde the *vendange* for the years 1838 to 1853 started on dates running from 14th September to 1st October. After the introduction of sulphuring the dates (1854 to 1863) were from 13th August to 15th September, an average gain of ten days. Of course it could have its disadvantages; if a treatment was given too near the harvest, sulphur would get into the wine and produce an off flavour. The problem of pesticide residues had already arisen! Though drawing the wine from the lees will get rid of this, yet, says M. Marès, there should be an interval of a month between the last treatment and harvest.

Fifteen to twenty kilos of sulphur per ha. were being applied two or three times a year, and this may well have helped in the battle against the phylloxera that was to come. The sulphur fell eventually to the soil and tended to increase its acidity, thus making the vineyard soils more acceptable to the new vines on the horizon.

Dusting with sulphur is the standard remedy against vine powdery mildew to this day. For instance, French agriculture used 45,000 tons of sulphur dust in 1954, nearly all of which would have been used against this disease. In 1968 Italy, having a larger area of vines, used 141,000 tons of sulphur, mostly for the same purpose.[65]

The attack of the powdery mildew is one of the reasons for the decline of wine-growing in Britain. Many old houses had vines on outer walls, which gave good crops of grapes and wine before the arrival of the mildew. As the disease spread the crop declined, reinforcing the erroneous idea that grapes cannot be grown out of doors in Britain. Badly mildewed vines can still be found on walls of many old houses and gardens, and these plants can usually be restored to health by spraying or dusting with some form of sulphur.

The disease not only reduces the crop very considerably, but the grapes that are picked give a wine of very poor quality, suitable only for distilling.

The effect of this disease may be seen from the series of production statistics for the Gironde, running back to 1840. These figures show that before sulphuring became general the powdery mildew took between half and two-thirds of the crop, but even so the oidium was but a rehearsal for the phylloxera. Amateur actors whose dress rehearsal goes badly console themselves with the thought that 'it will be all right on the night'. Alas, in this case the 'night' of the phylloxera was considerably worse.

Treatment with sulphur saved the vines and showed that pests could be conquered by the application of science and common sense. The general excitement caused by oidium may be seen by the fact that scientists and the press were inundated with a flood of suggestions for methods of control. As early as 1853 a M. Victor Rendu comments on the matter:

> I cannot even start on the formidable list of suggestions put forward as cures; they make a fine catalogue . . . what crack-pot inventions,

what highly praised cures! Nothing else could be expected when a disease shows such contrasting facts and combinations . . . when its cause is unknown, empiricism steps in. It is hoped that truth will be found by some chance discovery; hence the flood of suggestions that nearly drowned us. . . . Unfortunately good faith is not always free from exaggeration, and all too often the infallible success, prematurely announced, leads to failure. The story of the vine mildew is but another proof of this sad truth.[151]

It was a story that repeated itself.

CHAPTER FOUR
America

America was probably the home of the powdery mildew; it certainly was that of the vine phylloxera. Scientifically *Phylloxera* is the name of a genus of aphids, and species of this genus found on other plants, such as the oak, were known from 1834 onwards —the date of the establishment of the genus by Fonscolombe.[77] The story of the vine pest phylloxera starts in 1856 when an American entomologist, Dr Asa Fitch, described an insect from vine leaves and named it *Pemphigus vitifolii*.[70] Ten years later three other American entomologists, C. V. Riley, about whom we shall have much to say, H. Shimer and B. D. Walsh, examined the creature and found it had been wrongly placed in the genus *Pemphigus*. There was considerable discussion about its name and classification; Mr Walsh maintained that the creature was a coccid—that is, a scale insect—and Dr Shimer that it was an aphid warranting the establishment of a new family, which he proposed calling the Dactylosphaeridae. Mr Riley agreed that it was an aphid, but not one that needed a new family. It was purely and simply a species of Fonscolombe's *Phylloxera*.

The early white settlers in America, particularly the French, brought *vinifera* vines with them and tried to establish vineyards of the varieties which were so successful at home, yet always were they disappointed. If the grower were lucky, the plantation might last a few years, but sooner or later it would slowly sicken and die.

French colonists in America are said to have planted European vines in Florida as early as 1564, and by 1630 there were English plantings in Virginia. They started off so well that in the same year the London Company sent French vineyard workers out to look after the grapes. The plantations then started to fade away, and these unfortunate men were accused of neglecting the vineyards. There is no doubt that the plants were being killed by the phylloxera. In 1635 William Penn tried, and failed, to establish a vineyard in Pennsylvania, and in 1690 a Swiss colony attempted to reproduce the splendid wines of the Lake Leman area in Jessamine County, Kentucky. Their first capital of $10,000, a considerable sum in those days, was lost and they moved to Vevay, Indiana, and grew a native vine called Cape or Schuylkill. It seems to have faded away, and by 1819 the former vineyards were wheat fields. The same story is told over and over again. Pierre Legaud, from Lorraine, repeatedly tried to grow French, Spanish and Portuguese vines and repeatedly failed. Two Frenchmen, old soldiers of Napoleon driven out of Texas, then part of Mexico, tried to establish *vinifera* vineyards at Tombig Bee River, Marenga, Alabama, one of them being no other than the noted M. Lakanal, a member of the French National Convention who, during the reign of that body (1792–5), founded the Institut de France and the Natural History Museum, Paris. Similar fates with vines befell planters in Tennessee and Ohio. It was thought, erroneously, that the climate and soil were so different that the European vine would not thrive, a difficult proposition to accept, because so many European plants throve to excess in America. It was surprising that none of these people noticed the aphids on the roots, and we shall come back to this point later.

In view of the continued failure of the imported varieties a number of farmers started to grow the native grapes, but the wine they made does not seem to have appealed to the colonists very much and no large plantations of these were made in the eighteenth century. However, by the beginning of the next century American grapes began to come into their own and were thriving by the mid century. A Swiss immigrant at New Vevay, on the banks of the Ohio River, and Lakanal, now a congressman, and others in Kentucky, Tennessee, Ohio and Alabama are known to have failed with European vines and then to have turned to *labrusca*, an obvious choice because of its large and prolific fruit. This was followed by plantings of *riparia* and *æstivalis*, and in the south by *rotundifolia*.

There was a considerable interest in these plants and hybrids were developed. The pioneer grower of American vines on a large scale was Mr Longworth of Ohio, who started in 1823 and was followed by many others, among whom may be mentioned Messrs Adlum, Allen, Arnold, Bull, Roger and Underhill. A considerable demand began to build up, both for fresh grapes and for the wine made from them. It must be remembered that not all American species are immune from the insect. Some of these vineyards were attacked and considerable damage done. Various wild theories were put forward as to the cause, but it was not until the 1870s that much was done about it, mostly because it was mysterious and there appeared to be nothing that *could* be done. The *laissez-faire* philosophy of the rapidly developing U.S.A. was not conducive to the spending of taxes on scientific research. It was only after news had been received from France of the losses the pest was causing there that the great C. V. Riley turned his attention to the matter.

In the case of the American vineyards we can see two things happening, first, how the insect shaped the industry, and secondly how empiricism cured a trouble when it was not known that that particular trouble existed—like being cured of a disease one did not know one had. But the French grape-growers in Louisiana in the late seventeenth century had a worse trouble to face than the phylloxera—French home policy and decrees. The home government, wishing to protect its wine export trade, obliged the colonists to pull up their vineyards and many were destroyed, including the then thriving plantation of the Jesuits at Kaskaskia, Illinois, no doubt one of the reasons why the colony did not resent being sold to the U.S.A. in 1803—the 'Louisiana purchase'. It is ironic to think that the French Government need not have worried; they could have avoided the obloquy by letting the phylloxera do the dirty work for them. In fact they really did the colonists a good turn, for they saved them from much slow, inevitable disappointment with vines by forcing them to turn to other crops.

The banks of the Ohio River were later to become a great vineyard area. In July 1796 the erudite French savant and writer Volney visited some French settlers in the area, and at Gallipolis tasted a red wine said to be made from European grapes, but which the expert Mr Buchanan said were really *labrusca*. M.

Volney was not very impressed with the vintage and stigmatized it as 'a nasty little Suresnes', the latter being a suburb of Paris, presumably notorious for its sophistication (adulteration) of wines at that time or for bad wine from mildewed grapes.

Around 1805 John-James Dufour, living at Vevey, Switzerland, began to take an interest in wine-growing in America. He had read in Swiss newspapers that the French officers serving under Lafayette in the revolutionary war complained that there was no wine to be had in America. Dufour was a temperance advocate, which ideal he thought could best be promoted by substituting good wine for the heavy spirits causing so much drunkenness. He argued that vines surely must flourish in America at the same latitude as they did in Europe; after all, the European cereals did, why not the vines? So enthusiastic did Dufour become that he went to America. After three years of searching he found a suitable spot, a small plantation of European vines owned by a Frenchman named Legaux at Spring Mill on the Schuylkill River. Legaux's vines were said to have come from the famous Constantia vineyards, South Africa, and were thriving, possibly because he had been lucky and no phylloxera had reached his isolated plantation, or because he really had a more resistant American species or hybrid and not the Constantia *vinifera*. Dufour planted a vineyard in the Great Bend of the Kentucky River some twenty miles from Lexington, using cuttings from Legaux's vineyard and later some 10,000 cuttings, comprising thirty-four varieties of *vinifera*, from Europe. In 1798 he founded the Kentucky Wine Society, and three years later he brought in seventeen members of his family to help run the vineyards. He also wrote a book on the subject.[60] All this activity seems to have led to the introduction of the phylloxera, and the vineyards slowly failed. The whole Swiss colony then moved to a new site on the Ohio in Indiana and called it New Vevay. They planted the Schuylkill grape, called after the river of that name on which Legaux had established himself, another fact suggesting that Legaux really had an American species and not the 'Cape' he thought he had. Foëx and Vialla [76] classify Schuylkill as of 'unknown origin'. It appears to have been discovered by Governor Penn's gardener (Mr Alexander) in Pennsylvania before the War of Independence. The Schuylkill wine was likened to a red Bordeaux and was the only American wine of any repute until the Catawba came along. In

1818 Isabella was found, named after a lady called Isabella Gibbs in South Carolina (it is often grown today in the tropics). It too was a *labrusca*, a species not particularly resistant to the pest. Isabella slowly faded away, and we may note that the same thing happened in France when Laliman planted it in his experimental vineyard.

Major Adlum introduced the famous Catawba in 1818, having found it in the garden of a German family near Washington, though there is a tradition that it had come from Buncombe County, North Carolina, and in fact it bears the name of its supposed place of origin—the banks of the Catawba River. It too was a *labrusca*, and it will be seen what great reliance was being put on this species at that time, but Catawba was rather more vigorous than most *labruscas* and thus tended to resist the phylloxera for a longer period. Moreover it had a better flavour, being less foxy than the other cultivars of this species. In the 1870s it was 'the pride of the United States', and the major declared that in introducing it he had 'done the nation more good than if he had paid the national debt'. No doubt the gallant soldier did himself a bit of good too selling plants, let alone the wine. One of Major Adlum's customers was Nicholas Longworth, born in New Jersey in 1803. He became a lawyer, settled in what was then a village of log huts and 800 inhabitants called Cincinnati, and became very rich by buying and selling land. A mere twenty-five years old, he retired from the real estate business in 1828 to take up wine-growing and turned the Ohio into the 'Rhine of America', helped by a large influx of German immigrants who knew viticulture. Having failed with 1,500 European plants Longworth turned to Adlum's Catawba, his efforts being assisted by Major Adlum's book on the cultivation of the vine [2] (1823), a work based mostly on English authors.

By 1845 there were eighty-two vineyards covering an area of 122 ha.; this had grown to about 500 ha. by 1852 and 4,600 by 1870 in Ohio alone.

Longfellow sang the vintage's praises in his poem 'Catawba Wine', and one must suppose either that Major Adlum and his successors had somehow suppressed the foxy taste, or that the poet had acquired a liking for its strong flavour. We may note that at this time the American press was making much of the supposed adulteration of French wine.

. . .

Very good in its way,
Is the Verzenay
Or the Sillery soft and creamy
But Catawba wine
Has a taste more divine,
More dulcet, delicious and dreamy.

There grows no vine
By the haunted Rhine
By Danube or Guadalquivir,
Nor on island or cape
That bears such a grape
As grows by the Beautiful River.*

Drugged in their juice
For foreign use,
When shipped o'er the reeling Atlantic,
To rack our brains
With the fever pains
That have driven the Old World frantic.

. . .

And this Song of the Vine
This greeting of mine,
The winds and the birds shall deliver
To the Queen of the West,
In her garlands dressed
On the banks of the Beautiful River.

Not, one must admit, one of the poet's happiest efforts, but no doubt sincere.

By the 1870s Catawba cultivation was declining, due to black rot, mildew and phylloxera, and Concord, still a *labrusca*, was taking its place, a vigorous fertile variety more resistant to pests, but also more foxy-tasting than Catawba. It was a described as 'the grape for the million'.

New varieties would be 'discovered' from time to time, and the *æstivalis* species came into use. It was much more resistant to the insect than *labrusca* and throve where *labrusca* failed. In the

* Presumably the meaning of the Indian word 'Catawba'.

1830s Herbemont, an *æstivalis*, was regarded as promising and was said to have been found by Nicholas Herbemont in South Carolina in 1798. It was also called Warren, having been discovered growing wild in Warren County, Georgia. Cunningham (*æstivalis*) was a variety put forward by Jacob Cunningham in Prince Edward County, Virginia, and made into wine by Dr Norton in 1835. It too was vigorous and was called the 'bags of wine' grape by its protagonists. Herbemont (Warren) and Cunningham proved to be the most vigorous vine in tests subsequently made at Montpellier, France.

In 1848 the first plantings were made at Hermann, near St Louis, using Isabella and Catawba, but these varieties slowly failed; nevertheless the vineyards spread along the Pacific and Missouri railway line as it moved westward, *æstivalis* varieties, such as Norton's Virginia and Clinton, restoring the faith of the *vignerons*, particularly those along the banks of the Mississippi River. About 1848 Catawba was being planted at Kelly Island, Lake Erie, by Mr James Carpenter, a distinguished farmer.

Some figures collected by Mr Isidor Bush [36] show the area and yield of that distant time. The year 1840 saw the production of 4,720 hl. of wine. By 1858 some 2,450 ha. in eight mid-eastern states produced 75,700 hl. and Ohio had just under half this new vine area. By 1860 the production was down to 70,400 hl. possibly due to the phylloxera, bad weather or both. In 1871 about 530,000 hl. were secured. For comparison, the 1967 U.S.A. production was over 11 million hl.[65]

Wine-growing in California was quite a different story. There the European vines prospered until the 1870s; there were no mysterious, slow failures, nor did the indigenous varieties suffer. The reason was that the phylloxera did not exist west of the Rockies up to that time.

At Kelly Island, Lake Erie, in 1860 Mr Thomas Rush, a German immigrant, who presumably had anglicized his name, planted 800 vines he brought from Neustadt-an-der-Haardt, Bavaria. After three years they began to fail, and as they did so the owner had the sense to replace them with local varieties. In the 1870s the produce of this vineyard became the best-known American white wine, due not only to a happy selection of grape varieties but also to skill in manufacture whereby the foxy taste of the *labruscas* was much reduced. Planchon [141] has an interesting description of the winery he found at the Kelly Island Wine Company during his visit there in 1873:

> It was a vast building in dressed stone, heavy in appearance and with four round towers at the corners giving it the false appearance of a feudal castle. The above ground portion was a room 58 metres long by 25 wide. Here the pressing and fermentation took place. Two wooden floors divided the building into three stories; on the ground floor there were six big presses. The grapes were brought in from the country by various growers, tipped into a large box running on rails and wheeled to the weighbridge, when payment was at once made for them. They were off-loaded into a large receptacle from which an elevator raised them to the hopper of the de-stemming machine on the second floor. This machine extracted the stems and allowed only juice and grapes to go forward. The juice ran straight to the fermenting vats on the first floor and the pulp, pips and skins went to the presses on the ground floor * which were worked by a 15 h.p. steam-engine, though they could also be worked by hand.

Each press treated three tons of grapes and took six hours to do it; so quickly did they work and so good was the organization that seventy-two tons of grapes could be processed in a day, suggesting that they were working twenty-four hours a day, no mean feat in the days before electric light. There were at least nine other wineries on the island, and a number of other Lake Erie islands, such as Middle Bass, also had them.

What a tremendous enterprise all this was and how busy they must have been during the vintage! How did they ensure a steady flow of grapes to the winery night and day? There were no telephones to stop a man coming in the case of a glut or to urge him to come to fill a gap. There was probably a lot of waiting about, and one advantage the driver or wine-farmer had over the moderns was that if he had to wait, or in any case after he had delivered his load and been paid, he could refresh himself on wine to his heart's content. He had but to point his horse's head in the right direc-

* Note difference between European and U.S.A. usage. In most of Europe the first floor is the floor above the ground floor. In the U.S.A. the ground floor is called the first floor. Planchon probably did not know this, or the distinction may not then have existed.

tion and the animal would take him home with a minimum of guidance, a facility not yet built into the motor vehicle.

The real solution to the 'decline' problem was to try to find a hybrid combining the quality of the desired grape, were it a *vinifera* or strong *labrusca*, with a resistance or immunity to phylloxera, and it is surprising how early either natural or deliberately produced hybrids were used. In 1806 Bernard M'Mahon [117] described fifty-five varieties of grape of which four were American vines; three of these were *vinifera* × *labrusca* seedlings. M'Mahon also made the suggestion about this time (though he was not the first to do so, as we shall see below) that was eventually to save the industry: 'I would suggest the idea of grafting some of the best European kinds on our most vigorous vines which no doubt would answer a very good purpose.'

In addition to the three books already mentioned there were several others, such as those of Rafinesque,[148] Downing,[58] Buchanan,[31] Le Conte,[45] Husmann,[93 94] Strong,[170] and Culler.[49]

With this very great interest in grape culture, with all these books on the subject, with European vines slowly and inevitably fading away and even with naturalists such as Lakanal interested in the subject, it is truly remarkable that no one ever noticed the phylloxera aphids on the roots of the vines. It might have been difficult to accept the fact that the aphid was causing the trouble, but that it was never even reported as existing on the roots was most strange. Surely those unfortunate Frenchmen brought out by the London Company to Virginia in 1630 and blamed for neglecting the vines must have seen the *vinifera* roots swarming with aphids and, grasping at any pretext to save their jobs and reputation, have blamed these insects for the trouble however unlikely a story it may have seemed? Of course this is hindsight; it now seems so obvious and it bears a lesson for us too, mostly that we should beware of too readily accepting the *idées reçues* of our age until they have been tested. In the nineteenth century the idea prevailed that the soils and climate of North America did not suit the European vine, a situation not unacceptable to European governments doing a reasonably good export trade in wine, for it automatically killed competition. Aphids leave dying vines; by the time the farmer comes along and digs up a dead, or nearly dead, vine there are no aphids left on the roots, and it is difficult to associate cause and effect. An even greater difficulty would have been to believe that such a tiny creature as a phylloxera aphid could have such devastating effects. Not even the great Riley reported the root aphids until his attention was drawn to them by the events in France.

Charles Valentine Riley (*see* Pl. 8) was a remarkable man about whom we need to know more, for he played a vital part in overcoming this pest. He was born at Caroline Cottage, Queen's Street, Chelsea, London, on 18th September 1843. His father was Charles Riley, a commercial traveller, and his mother before marriage was Miss Mary Louisa Cannon. The family moved to Walton-on-Thames; nevertheless Riley went to school in Chelsea and Bayswater. At the age of eleven he went to a military school, the Collège de St Paul, Dieppe, and three years later to Bonn, where he again spent three years, thus in six years acquiring considerable fluency in French and German. From an early age he was interested in insects and showed great skill in drawing them. At the age of seventeen he lost his father and had a younger brother to support. In 1860 Riley literally sailed to New York, a voyage taking seven weeks, as an immigrant, and a year or so later his mother in England married again, a gentleman named Lafarge. Young Riley obtained a job on a stock farm in Kankakee County, Illinois, where he stayed for three years, working and studying with considerable success. It proved too much for him and his health broke down; after recovering he went to Chicago where he experienced very hard times, though he eked out an existence writing on insects for the *Prairie Farmer*.

In 1863 the Colorado beetle reached Illinois and Riley described it very effectively. The farmers were beginning to realize the importance of knowledge in dealing with insect pests, and the *Prairie Farmer* appointed Riley the editor of its entomological department in 1864. Riley then entered into an extensive correspondence with B. D. Walsh, the acting entomologist of the state of Illinois, which demonstrated his sound grasp of the fundamentals of his trade, though he was but twenty-one years old. At this time the Civil War was raging and Riley joined the 134th Illinois Volunteers as a private in May 1864, the regiment being disbanded in November the same year (*see* Note 3). In 1866 he became entomologist for Illinois, a post created for the first time for him and which he held for nine years. It was money exceptionally well spent, for Riley's investigations of insect pests, and his remedies for countering the losses they caused, saved the state in particular, and the world in general, many millions of dollars.

During this period Riley wrote extensively, his most famous productions being the Missouri annual reports, admired for their accuracy, novelty and effective presentation.[154] Riley created a special style for them, for he realized that to be effective they must be read, and to be read they must appeal to the practical farmer. Moreover he had to overcome the suspicion that he, an immigrant, was trying to tell established citizens just what they should do. He conquered these difficulties by relegating the strictly scientific details to footnotes and elaborate appendices, usually printed in such small type that the average farmer would be unable to read them in any case; then he would enliven his account of a pest with a fresh style and mix in quotations, poetry and anecdotes, always keeping the main aim in view—to explain how a pest lived and how it could be overcome. He realized the importance of catching the reader's attention from the first, and most of his articles start with a striking and memorable sentence. For instance, his first account of the phylloxera [154] starts 'Here we have an insect, the life-history of which is as interesting to the entomologist as its devastations are alarming to the grape-grower,' a splendid example of the balanced, tripping, condensed sentence. In the same article he misquotes Shakespeare,

> ''Tis better far, to bear those ills we have
> Than fly to others that we know not of.'

In this particular instance Riley did not agree with the poet. He was also fond of quoting Tennyson and of introducing analogies from the Greek and Latin classics, though he did not lard his articles with quotations in those tongues.

These famous reports were admired by all and rapidly went out of print. Charles Darwin liked them and said he was struck with admiration at the author's powers of observation. Riley, in his turn, was slow to accept the principles of Darwinism, but did so eventually. His mind was far too logical to reject them altogether; he was not averse from admitting that he had been wrong on other occasions as well. One of his first triumphs was his investigation of the life of the 'seventeen-year locust' (*Cicada septemdecem* Linn.). He noted that there were both a thirteen-year cycle and a seventeen-year one and that the adult stage of both cycles coincided in 1868, leading to an enormous invasion. The locusts coinciding in this way greatly struck his imagination '. . . such an event has not taken place since the year 1647, nor will it take place again till the year 2089'. And the poet in him speculates on this event having occurred regularly for centuries in the past 'long 'ere Columbus trod on American soil'. Riley gave much good advice on the control of this pest; he not only wrote these reports but also illustrated them himself with very effective woodcuts.

At the age of thirty-five, in 1878, Riley was appointed entomologist to the United States Department of Agriculture at Washington, D.C., a position he lost when the government changed in May 1879. He returned to his post in 1881. From 1869 onwards he made several visits to Europe to study the phylloxera. He also conducted various French scientists around some of the vineyards of the U.S.A. to see for themselves how the phylloxera lived. M. Signoret, besieged in Paris by the Germans the previous year, had continued to study the phylloxera, writing to Riley by the balloon post.

By 1894 Riley was in poor health and resigned his position, although he retained the honorary curatorship of the department of insects in the U.S. Museum at Washington. In 1884 he had given his collection of some 115,000 specimens, representing more than 15,000 species, to the nation. During the last few years of his life Riley was entomologist and physiologist to the Maryland Agriculture Experiment Station, now at Beltsville, the world's largest institution of this nature.

Riley married Miss Emilie Conzelman in St Louis in 1878 and had seven children (*see* Note 4). On the morning of 14th September 1895 he was bicycling to Washington with his eldest son when his front wheel hit a loose stone. He was thrown from the machine and received injuries from which he died, aged just under fifty-two years.

As we shall see, Riley greatly helped France to overcome phylloxera. Not only the French Government recognized this by presenting him with a gold medal, but the French winegrowers and the firm of Vermorel as well, for they gave him a bronze statue commemorating the victory over the pest; it is now on display in the lobby of the Entomological Society of

Washington, D.C. (*see* Pl. 9). Riley's help was not confined solely to the phylloxera, for he made a very practical contribution to the control of the next trouble to strike the unfortunate wine-growers —the downy mildew—a disease, also from America, controlled by spraying with 'Bordeaux mixture'. Riley's contribution to this was the eddy chamber nozzle—also known as the Riley nozzle— familiar to anyone who has sprayed plants. The spray liquid,

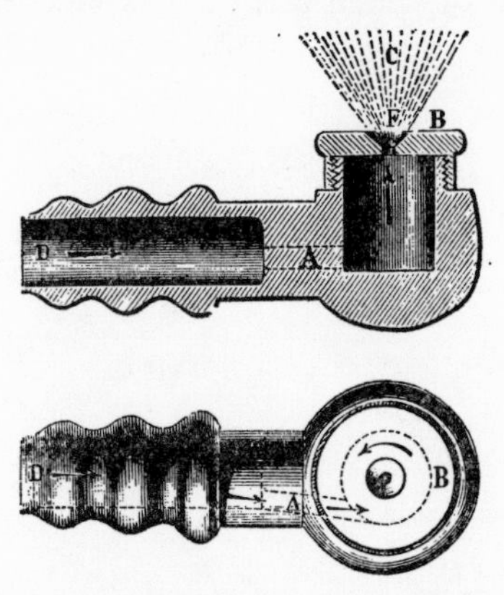

2. *The Riley nozzle, the device that made effective spraying possible against the downy mildew. The liquid whirls round inside the eddy chamber and emerges as a hollow cone of fine drops.*

under pressure, enters the round eddy chamber of the nozzle, say, of half an inch diameter, at a tangent and thus whirls round it, escaping from a central hole as a hollow cone of liquid quickly breaking up into fine droplets (*see* Fig. 2). This principle is now almost universal in crop spraying machines and few realize that it was C. V. Riley who invented it. French textbooks of that age refer quite simply to *un Riley* and everyone knew what it meant.*

Riley so much assisted the world to conquer pests (the Colorado beetle and the cottony cushion scale among many others) that his memory lingers on. As late as 1956 the two Misses Riley, his daughters, were walking in a vineyard near Bordeaux when they

* P. Viala (1887) [178] refers to the nozzle of the Noël sprayer as: '*C'est une boîte de* Riley, *fermée par un bouchon à orifice de grand diametre . . . semblable à celle d'un* Riley *ordinaire.*'

were stopped and asked if, in fact, they were the daughters of the great American who had saved the vines. On saying they were they were congratulated and much fêted. Needless to say the Misses Riley spoke French.

The United States eventually came in for a considerable amount of abuse for having 'sent' the phylloxera to Europe, but the account was set to right by the fact that they also provided the cure, a major part of which was Charles Valentine Riley himself.

CHAPTER FIVE
Unknown to Known

We have already noted that in the late 1860s an unknown disease was reported as causing considerable damage in the French vineyards. It has been generally accepted that the first printed report is a letter dated 8th November 1867 from a veterinarian, M. Delorme [54] of Arles, a town in the Bouches-du-Rhône. But in fact there are earlier references in the agricultural press. M. Delorme's letter stands out because it is the first accurate description of the new trouble, so we will deal with that first and return to the other, vaguer accounts later. M. Delorme's letter is addressed to the President of the Agricultural Show Society (Comice Agricole) of Aix, said to have been a very active man. M. Delorme talks of a vineyard planted in 1863 at Saint-Martin-de-Crau, between Arles and Salon, which gave a good crop in 1865. M. Delorme must have been a good observer, and he expressed himself in this letter so well about the symptoms that it is well worth quoting, for they are as true today as they were then, except possibly the speed with which the trouble spread. M. Delorme says:

> At the end of July 1866 the wine growers noticed a number of plants where the leaves had lost their normal dark green colour, some of them taking on a reddish tinge. The trouble spread outwards from the first attacked plants, that is from north to south, over four or five rows, and all the affected plants were near each other. The leaves quickly turned dark red and by the end of August

> every one of them had fallen. At this time there were about 200 affected plants. . . . When pruning started at the beginning of December, most of the plants in this patch were dry and brittle in their upper parts and some were quite dead. It then occurred to me to dig up one of the plants whose ground parts seemed full of sap at that moment to see what the roots were like. To my great surprise I found these almost as bad as the extension growths above ground, as many of the roots were already dead. On some a slight pressure between thumb and finger was enough to detach the surface skin. The root tissue was dark and the roots broke as easily as did the dry wood.
>
> By the end of February 1867 all the affected plants were dead. During this period and for the whole winter the disease had not stopped spreading in every direction, always from an affected vine to its neighbour. It was also just the same during the summer. In September, at harvest time, there were about five hectares of dead or dying vines and the crop was almost nothing.

The agricultural papers of that time contain many references to the new trouble. For instance, *Le Journal d'Agriculture Pratique* in November 1867 [100a] is reporting the summer discovery of a new vine disease at Narbonne, where a proprietor (M. Peyras) has asked the Comice Agricole to appoint a commission to examine it. Such a commission had been appointed in June, under a M. Martin-Donos, who found 'spots on the leaves, which spots then spread, the leaves fall and the plant dies', so the 'new disease' may not necessarily have been the phylloxera. Most likely M. Peyras's vineyard was attacked both by the phylloxera and the anthracnose fungus. That same year (1867) in the Gard and Vaucluse M. de Penanrun [137] reports on the new trouble, known for the last three seasons. It 'sleeps in the winter and wakens again in the spring'. Various growers got together to discuss the matter and they brought a dead, a dying and a healthy vine to the Comice Agricole, but again the aphids were not noticed. Perhaps the 'dying vine' was already fairly far gone and the aphids had already left it. The general consensus of opinion was that the vineyards were showing soil exhaustion.

The following year (1868) a summer report on the vines from Roquemaure, Gard, says that it cannot end on good news because in May M. Joulie had recorded a 'fungus disease worse than the

Oidium and he thinks it is the "black disease" noticed by M. Dunal, of Montpellier, in 1844'.

In all probability the first actual attack was at Pujualt, in the Gard, a little to the north of Arles, as early as 1863, for, as we have noted, M. de Penanrun of Ville-les-Avignons talks of an earlier attack and of finding mycelium growing on the roots of the dead vines. Naturally they thought that this was likely to be the cause and that it was due to a fungus known as the rot, or *blanquet*. Tree-root rot, or 'bootlace disease', is not uncommon in vineyards planted on old forest sites. It is caused by the honey fungus (*Armillaria melleus*) which sometimes spreads by sending thick fungus threads through the soil, not unlike bootlaces, and eventually kills the plant it attacks, often sending up a bunch of toadstools, which, incidentally, are edible, but not particularly good and in any case a poor compensation for the loss of a tree or vine.

These attacks of *blanquet* seemed strange to the *vignerons*, for it was the young vines on warm, well-drained slopes that seemed to be suffering most instead of those on the wet heavy lands where this trouble was usually found.

Planchon later (1877) [142] came to the conclusion that a considerable import of rooted American vines had been made between 1858 and 1862, 'by a singular coincidence' as he puts it, and that they had been sent to many parts of Europe, as far apart as Bordeaux, Roquemaure, England, Ireland, Alsace, Germany and Portugal. We have already mentioned the import trade to Britain of living plants and broadly similar figures are available for France. They run from 460 tons, worth 230,000 fr., in 1865 to about 2,000 tons in the 1890s, worth around 2 million fr. Thus considerable quantities of plants were being brought into France annually with no thought of pest transference arising: in 1875 nearly 50 tons came from the U.S.A., probably most of it vines.

The *blanquet*, *étisie* or unknown disease spread and caused considerable alarm. The rehearsal—the powdery mildew attack—was only too well remembered by the wine-growers, consequently the Vaucluse agricultural society and M. Gautier, Mayor of St-Rémy (Bouches-du-Rhône), got together and asked the Société d'Agriculture de l'Hérault for help.

A commission consisting of MM. G. Bazille, J.-E. Planchon

and F. Sahut was quickly formed, and on 15th July 1868, full of determination and energy (they would have had no hangovers for 14th July was not a fête under Louis Napoleon's régime) these gentlemen assembled at the Château de Lagoy, near St-Rémy, to investigate the matter. Up to this moment farmers had dug up dead vines and carefully examined them. They found no aphids because as soon as the sap pressure of a plant falls aphids feeding on it withdraw their feeding tubes and move off to seek a better food source. Consequently dead vines never showed the aphids, only the gnarled and rotting roots attacked by the saprophytic fungi. One did not dig up a sick vine as there was always the hope that it would get better, and in fact this could happen at times. *Vinifera* vines occasionally re-established themselves if they were very vigorous and the insect happened to be a little restrained for some reason. It was a struggle determined by the relative vigour of pest and plant.

M. J.-E. Planchon, the professor of pharmacy at Montpellier University, was a particularly able man. All kinds of wild theories were being put forward to account for the unknown disease which could not just be ignored but needed examination. Dr Watson tells us that when Sherlock Holmes was completely baffled by a mystery he consulted his brother Mycroft, a well-established, plump civil servant with greater deductive powers than those of the great detective. Mycroft would sit in his chair, and without moving from it solve mysteries, by pure reason. Similarly Planchon, sitting in his laboratory and thinking about the few known and verified facts of the mysterious new disease, decided from the way it spread that the cause must be something organic. The attacked area extended like a drop of oil spreading over the surface of a bucket of water. He dismissed the possibilities of the cause being over-production, bad weather or divine wrath; it had to be something living, organic.

The Commission consisted of trained scientists, and we can picture them that sunny, hot July day, somewhat formally dressed if we can believe the engravings in the press (*see* Pl. 5), accompanying the worried manager of the property. 'Gentlemen,' he said, 'it is disastrous: it goes forward like an army, laying waste all before it.' [42] Several workmen with mattocks, spades, baskets and argument were with them. Following behind would be some women with a few good bottles and a *casse-croute* or so, whilst the neighbours, pretending not to be interested, would be watching out of the corners of their eyes to see what was happening. MM. Bazille, Planchon and Sahut were not content just to dig up dead vines; they attacked the whole gamut from healthy plants, and apparently healthy ones, to the dying and the bare, gaunt dead vines that were so distressing. As they moved around out would come the notebooks and down would go their observations in neat, spidery pencil writing, with much mopping of their brows, sweating under their top hats. M. Planchon soon noticed that all the dying vines had small yellow insects on their roots. On some the insects were so numerous that the roots appeared to be varnished yellow, and a hand lens soon showed them to be a species of aphid. However, the Commission jumped to no hasty conclusions; they worked that day and all the next and agreed that the insects were connected with the trouble, in fact its cause. We may here quote Planchon's own account:

> From that moment one fact of capital importance was established, namely that an almost invisible insect, developing underground by myriads of individuals, could bring about the destruction of even the most vigorous vines.
>
> But what of this insect? From whence did it come? Had it been described? And what were its nearest relatives? These questions were not easy to answer there and then: in fact they could only be answered at all if all stages of the insect were found.[144]

Planchon gave the creature the provisional name of *Rhizaphis* ('root aphis', from the Greek), provisional because he felt sure that winged forms must exist and that these would establish its true zoological position. In effect he soon did find some aphids in the 'nymph' stage, showing wing buds, and he carefully preserved and watched them. On 28th August 1868 one of them hatched into 'an elegant little aphid with four flat, transparent wings'. Although he was professor of pharmacy, Planchon was also an entomologist and he immediately recognized the insect as being very similar to an insect living on the underside of oak leaves, *Phylloxera quercus*. Obviously it belonged to the same genus, and he named it *Phylloxera vastatrix*, thus declaring to all the world his belief in its being the cause of the trouble. So much education being based on the classics in those days, far more of the educated would have been aware of the meaning of *vastatrix* (and of *Phylloxera*, come to that) than would be the case today. It was the devastating dry leaf creature. The thing to do now, Planchon realized, was to work out the insect's life history, find the weak link, attack at that point and save wine and the vine for the world.

What talk there was at the Château Lagoy and in the village that night, the community splitting into two groups, one dazzled by the possibilities of science, the men with their apparatus, microscopes and machines that could solve anything, and the other feeling that such a set of impractical *savants*, who had never in their lives followed a plough or turned a wine press, could not really be trusted to overcome this problem.

It must be remembered that this was 1868, before the biologists had had any great triumphs in the agricultural or medical fields. When Pasteur, a few years later, showed that the *pébrine* disease of silkworms could be controlled and that rabies could be cured scientists gained tremendous prestige and things were different. This attitude persisted until the 1940s, it tending to be held generally that a man who knew all about, say, compression physics was entitled to pontificate on politics, religion and anything else. After the explosion of the hydrogen bombs the world began to realize that outside their immediate field scientists could be as stupid as anyone else; inside it at times too.

CHAPTER SIX

Effect or Cause?

One of Von Humboldt's more cynical observations is his description of three stages through which any important scientific discovery passes in the public mind. These are (i) to doubt its existence, (ii) to deny its importance and (iii) to attribute the discovery to someone else. In the case of the phylloxera it could not be denied that something was wrong, but it *was* denied that a tiny insect was capable of killing a flourishing vine. As in the case of the potato blight, the true cause was held, by most of the general public, to be the effect, and a number of scientists even took this view too. Finally disputes arose as to who had actually discovered the insect and named it, showing Von Humboldt's three stages very neatly.

The Commission's news was given to the world in the following sequence:

(i) 3rd August 1868. A letter to the Institute of France from M. Planchon.

(ii) 24th August 1868. Paper read by MM. Planchon and Saint-Pierre, in the Commission's name, to the Hérault Agricultural Society.

(iii) 14th September 1868. Note to the Institute, establishing the insect as a species of *Phylloxera*, with remarks on its propagation below and above ground.

It became obvious that vine-growers were faced with a serious threat. The following year the Société des Agriculteurs de France formed a special commission which included MM. Bazille and Planchon, the former being the vice-president, to investigate the subject. Planchon was continuing his studies at Sorgues, Vaucluse, and examined some American vines of a variety known as *tinto* and found galls on the leaves very like those described by Asa Fitch as caused by *Pemphigus vitifolii*. A few days later M. Laliman found these same galls on some American vines near Bordeaux, many of which had phylloxera aphids on their roots. The name raises a curious supposition, for 'tinto' suggests it was a Mexican or Californian variety, both areas where there was no phylloxera. This 'tinto' may well have become infected in France.

Leo Laliman was a Bordelais who had been born in 1817; after a strange, short and dramatic military career (*see* Note 5) he devoted himself to economics and viticulture, becoming particularly interested in American vines. He published a book on them in 1860 [108] in which he maintained that by using certain American varieties the French vineyards would be immune to attacks of the powdery mildew, pointing to such American vines growing free from disease in the Luxembourg Gardens, Paris, from 1817 onwards and on his own estate since 1840. He could be suspected of having brought in the phylloxera, but if 1840 was his only importation he can be cleared of this suspicion, for the pest would have become generalized far sooner than 1870 had Laliman been the culprit. However, he was probably importing vines from America all the time, and suspicions of this nature hanging over his head may have much harmed him in later years. In any case no particular individual could be *blamed* for importing the pest. It was not the philosophy of the age to impose any such restraints on 'business' as a prohibition of imports on phytosanitary grounds, an attitude that was later to change when it was realized that pest control laws regulating imports could be a method of protecting home production without imposing customs prohibitions and thus giving rise to retaliation by the countries affected.

Planchon was working with another biologist, his brother-in-law Lichtenstein, and the idea struck them that the comparatively innocuous leaf aphid described by Asa Fitch as *Pemphigus vitifolii* was the same as the dangerous root form they had just discovered and had named *Phylloxera vastatrix*. The two brothers-in-law started some experiments showing that the gall form transferred to the root form, and C. V. Riley came especially from America in 1869 to confirm the identity of the two insects. According to the strict rules of priority the pest should be called *Phylloxera vitifolii*, using Fitch's specific name and, in fact, many scientists do use this name, but *P. vastatrix* is so widely known and used that it is also accepted.

In August 1868 these events did not cause much of a stir in the Paris press, which was more concerned with the forthcoming Great Exhibition and the politics of Louis Napoleon's 'liberal empire'. At times during July, August and September 1868 *Le Temps*, particularly in its agricultural supplement, reports on the splendid vintage coming forward, that the oidium is no threat but that Pyralid caterpillars are doing much damage. There is no mention of the 'new disease', but there is an observation showing that it might have been more widespread than was thought. It was reported that some *vignerons* were afraid of unripe grapes in the vineyards: fruit that does not fill or ripen is one of the symptoms of phylloxera attack. More specialized papers were less reticent. M. Jean Tapié of *Le Petit Journal* started an inquiry on the new disease.

Planchon, who did so much to conquer the pest, was a dedicated scientist. He was born at Ganges, l'Hérault, on 21st March 1823, the son of a devout and humble Protestant family. In 1844 he came to Kew Gardens and worked there until 1849. Later he took a degree in medicine and became Professor of Medicine and Pharmacy at Montpellier University. At Kew Gardens Planchon was employed as assistant to Sir William Hooker and was in charge of Sir William's personal herbarium; he was not on the official staff of the gardens. He must have been much liked by the Hookers and he too liked them. A number of letters in the archives [106] at Kew show Planchon's continuing friendship with both Sir William and his son (Joseph D. Hooker) and eventual successor there, and with other Kew notables, such as Sir William Thistleton-Dyer and George Bentham. In his letters to Sir William Planchon frequently inquires about Mlle Elizabeth; he hopes she has recovered from her cold and is able to enjoy her country walks; he notes with pleasure that the family are in Devonshire and will be back at Kew just before he returns. One feels there may well have been a *tendresse* between the young Jules-Emile and Miss Elizabeth. But the young Planchon later married a sister of the amateur entomologist J. Lichtenstein, and the two brothers-in-law eventually made a very good phylloxera team.

Planchon's letters to Sir William and others are largely on botanical matters, but they contain much family news and philosophical remarks on such subjects as God and politics. For instance, in 1857 he tells of the death of his first child, a boy, and on 2nd January 1858 he announces that he is the proud father of a 'Miss' two days old. On 1st January 1871 Planchon writes a letter which might well have been considered treasonable had it been opened in transit. He bemoans the 'horrible war in which the ineptitude of the Empire has left us. France is paying for having cruelly suffered a subtle corruption for so long. The Prussians, who after Sedan could have calmed the passions of the struggle, now show the world the horrible spectacle of civilization put to the service of barbarity.' The French always appear to equate themselves with civilization more closely than any other nation. More people were killed by the Commune and its suppression than in any battle of the war. One must not forget that Louis Napoleon sought the war, and that between 1785 and 1813 France had invaded Germany fourteen times.[92] But Planchon was supreme in his profession and kept Sir William Hooker well informed on the phylloxera situation with personal letters, samples of American vines (many of them still growing at Kew Gardens) and printed reports. They served Sir William very well, for he was able in due course to inform the Marquis of Salisbury, the Foreign Secretary, that the phylloxera was a serious threat and not just a pest that would come and go as some of his consuls and ministers in Europe had suggested to him.[106]

The Paris dailies do not seem to have reported the Commission's discovery, but local papers buzzed with it. In 1874 Planchon wrote an account of the matter in a political fortnightly [140] in order to publicize the threat and to clear up many misconceptions, and we shall come back to this account later.

The news appears to have passed unnoticed in the popular press in England, but the *Gardeners' Chronicle and Agricultural Gazette* picked it up at once. J. O. Westwood immediately recognized it as the insect that had been sent to him from Hammersmith in 1863; on 30th January 1869 he had a column and a half in this journal.[186] In 1867 Westwood had received specimens of the insect from both Cheshire and Ireland, so that it was remarkably well spread out over the British Isles. He gave a lecture on the subject to the Ashmolean Society at Oxford in the spring of 1869, which seems to have passed unnoticed in France. He made a lantern slide showing the insect, a gall and a galled leaf, which served as the basis of a woodcut for the *Gardeners' Chronicle* article of 30th January 1869. Though rather small, from the general shape of the leaf and the fact that the leaf has galls on it (which is rare in *vinifera*) it is quite possible that the infected leaf was from an American *Vitis* species. Westwood named the insect *Peritymbia Vitisana* 'in allusion to the tomb-like gall on the leaves formed by the female insect'. The specific name suggests the very opposite of what was to prove to be the case! Westwood then refers to the Planchon discovery and the naming of the creature as *Rhizaphis vastatrix* in the summer of 1868, 'a name', as M.J.B. (the Reverend M. J. Berkeley) well observed, that was 'scarcely applicable, should it turn out, as we suspect will be the case, to be congeneric with the very similar insect which is found in the excrescences on Vine leaves'. Next reported is Lichtenstein's account of the ravages of the pest, made to the Entomological Society of France on 12th August 1868. The *Gardeners' Chronicle* also reported this with what might be considered an ominous misprint. Lichtenstein said that the pest had destroyed the vines from Arles to Orange on the left bank of the Rhône; the English periodical managed to print 'Rhine'. The French were soon to lose this bank to the Germans, and the struggle gave the phylloxera a great opportunity to consolidate its hold.

The Hérault Commission's announcement, as we have noted, was by no means universally accepted. It was agreed that phylloxera aphids could be found on the roots of the dying vines, but were they the cause or the effect of the trouble? It is extraordinary how the human mind refuses to accept the obvious, for the same thing happened with the potato blight; the fungus found on dying leaves was declared to be the effect of the disease and not the cause, as the Reverend M. J. Berkeley [21] had stated. And the same causes for the trouble were put forward in both cases: bad weather, over production, winter cold and God's punishment for the vices of the age. That the phylloxera was merely an effect of the basic trouble was the view taken by a number of scientists, one of them being the distinguished entomologist Signoret, who gave a good paper on the insect to the French

Entomological Society in December 1869,[164] in which his subtitle was 'the supposed cause of the present vine disease'. Though he was wrong as to cause, yet he described the insect very well in most of its stages and pointed out that no new genus (Planchon's *Rhizaphis*) was called for, it was simply a new species of *Phylloxera*. His paper reviews the various insects found on the vine from the earliest ages until the 1860s, and then details the sundry recent discoveries of the phylloxera on vines and quotes the views of various writers. For instance, in 1861, in the Indre-et-Loire and in the Haute-Savoie, it was known as 'cottis' or 'nettle-head'. Dr Guyot thought that its presence was due to over-severe pruning.

In a continuation of this publication Signoret says that, after reading Vialla's report,* he is more than ever convinced that the aphid is not the cause of the trouble.

The evidence that Signoret puts forward for this view is not very convincing. He starts off by admitting that a considerable number of sucking insects on a plant could lower the resources of that plant by the quantity of sap they extract, but he saw the cause of the trouble as (i) the long drought, (ii) 'bad' growing and (iii) the unsuitable nature of the invaded vineyard soils. It is strange that such an acute observer, and an entomologist too, did not reason from a fact he must often have seen, that aphids leave a plant as soon as the sap pressure falls. If you cut a rose with some greenfly on it and put it on, say, a dark polished table you will soon see aphids moving off across the dark surface; they are seeking a plant with enough cell pressure in it to force the sap into their feeding tubes when they insert them. Signoret, and others, can be criticized for not realizing from first premises that the phylloxera would leave a root as soon as it started to die. He did point out that the aphids leave the roots well before these got rotten. He also notes that the moister sites are the least attacked by the pest and goes on to quote from Planchon and

* Note, in passing, that we must avoid confusing two men with similar names, both working on vine pests at that time. Firstly there is L. Vialla, the entomologist, and secondly Pierre Viala (with one l), the expert in fungus diseases. That this can easily be done may be seen from the fact that contemporary writers confuse them in one article, and even the spellings 'Valla' and 'Villa' are found.

Vialla cases where vines, particularly those in damp situations, were known to have aphids on their roots in one year and were still thriving the next. Of course this can happen. Although a deep-rooted vine, particularly where the deep soil is compact, sandy or wet, may show phylloxera on its upper roots, where the insects may have destroyed most of the fine roots in the upper system, it can continue to be nourished by its deep roots, where the soil is sufficiently compact to resist penetration by the pest. One gets a feeling that Signoret only half believed his own theory. Obviously he was fascinated by the genus, and he continued to work on the creature when Paris was besieged by the Germans.

Many wine-growers and much of the public found it difficult to believe the phylloxera/cause theory and plumped for the phylloxera/effect suggestion, much to the annoyance of most of the scientists. In 1868 M. A. de Ceris, of *Le Journal d'Agriculture Pratique*,[100b] was firmly on the phylloxera/effect side in his reporting of events and he was supported by M. Henry Marès (who seems to have favoured the English spelling of his name, although sometimes it is in French), who was the permanent secretary of the Hérault Agricultural Society and later became president of the Phylloxera Commission. Obviously there came a point later on when he changed his views, but it is likely that his was the dissenting voice in the Vaucluse, mentioned below. At least two growers went on record as agreeing with Henry Marès, and a M. Alphandry (Jeune) protests energetically against the suggestion that within a year all the Vaucluse vines will be lost.

Writing of these events in 1874 Planchon [140] shows he was much concerned with the apparent blindness of people, even scientists, and sought to find its cause by an analysis of the facts; it is instructive to follow his argument. Not only were there the phylloxera/effect and phylloxera/cause schools of thought, there was also a 'sort of compromise between these two opposing systems', and he sees M. Henry Marès as the main advocate of this last theory. The 'effect' people thought, we might say, sexually. They held that phylloxera was the result of continued vegetative reproduction, which weakened plants. To maintain health sexual reproduction, that is, the introduction of male vigour, was needed from time to time. Obviously it was difficult for many people in that age to realize that in many spheres of life there was no need for males at all.

The strange thing is that in many cases the facts seemed to support the theory of degeneration through continued vegetative reproduction. New potato varieties, for instance, would start off splendidly and fail after a few seasons even after careful selection of seed tubers each year. The truth was that the degeneration was caused by virus diseases, about which nothing was known in that age, not even that viruses existed. When plants are reproduced by vegetative means (cuttings, tubers, buds, etc.) any virus disease present is also passed on, but very few viruses are passed in the seed. A 'seed' potato tuber can contain a virus disease and produce a weak plant, but the true seed, from a potato flower, will not usually be infected, though it may well not come 'true' to its parents. There was thus some cause, even for scientists, to think that sexual reproduction was necessary, but they had the wrong reasons for it. Planchon pointed out that even were this true such a wholesale and sudden degeneration of the vine, always coinciding with the presence of the insect, was beyond all reason to expect. Other causes put forward were soil exhaustion and short pruning, to which the same arguments applied. Planchon says the only reason he gives such absurdities the honour of refutation is partly from a spirit of impartiality, but principally because the French public is unacquainted with scientific method and is therefore unable to use common sense, particularly in judging matters of natural history when these are presented by subtle arguments which, nevertheless, fly in the face of the facts. Rhetoric thus can convince not only the uninstructed but also the educated, and otherwise sensible, citizen.

Planchon pours out a great deal of scorn one way and the other, particularly on the notion of sudden, widespread degeneration. Degeneration in greenhouses, in the open air, only on *vinifera* vines, here, there and everywhere, spreading from one vine to another, or jumping kilometres at a time, and the degenerate vines always then getting the phylloxera. This, he says, is the subject of predestination (at that time a subject much discussed by the clerics) extended to the vegetable world. We may just as well say, he continues, that sheep eat the wolves and cabbages devour their caterpillars.

The phylloxera/effect theorists, he continued, are found mostly among the visionaries. Most of them, instead of noting the facts and reasoning from them, wrongly use an *a priori* reasoning and try to solve the mystery by finding connections with other, far-off troubles having nothing in common. Often they confuse the oidium with the phylloxera and propose to cure both at the same time. Planchon, seemingly a mild, intelligent man, liberal and to some extent a rebel, being a Protestant, blames the old, pedantic educational traditions still influencing the science of his day, a point also made elsewhere by Fabre, the entomologist, at this same time. This education gave rise to 'a nightmare of hidden causes darkening with its phantoms the search for the light of truth' and the prize of 20,000 francs (the first prize offered for a cure) brought forth such a mass of absurd suggestions that it would be laughable were it not a humiliating indictment of the educational system.

However, the 1869 Commission appointed by the Société des Agriculteurs de France was a model of tact and let the dissenters down gently. Its report was written by M. Vialla.[181] It describes the damage done, says the cause is the aphid and then proceeds to expose all other theories as false. Vialla pointed out, however, that bad weather, poor soil and bad management were contributory causes, though not the main cause. For instance, there was some reason for thinking that bad weather was the culprit. The winter of 1868 was very cold and dry; the summer was yet drier. The Rhône fell so low that it was blocked with sandbanks, and some areas, it is said, had had no rain for eighteen months. The Commissioners pointed out, though, that the disease had started in 1864 and 1865, and a bad season could hardly have a retroactive effect; also that cold kills quickly but drought only slowly. Nevertheless though drought was not the cause it naturally contributed to the death of vines. The scientists on this mission made an important observation that was not followed up to any great purpose. They said they did not know if the insects destroyed the root tissue simply by their sucking action or if they injected some irritant substance into the plant. Today we know the latter is the case, but we still have very little information as to the nature of this 'irritant substance'.

In Provence vines were often planted on old woodlands, and they could suffer in such sites from the honey fungus, hence farmers tended to regard the new disease as of fungus origin and called it by the same name, rot or white rot. The Commission dismissed this possibility somewhat dryly with the remark:[181]

'However, one could not go on believing in the existence of a fungus nobody ever saw.'

The Commission thus came down firmly and unanimously on the phylloxera/cause side, but a commission set up by the Department of Vaucluse did so by only five votes to four. The majority of the wine-growers took the same view, but by no means all. As to the general public, they were slow to hear about it in the first case and slower still to accept the insect as the cause. As late as 1875 *Le Journal Illustré* was able to announce that joy reigned in Provence; one could dance on the bridge at Avignon again. Yesterday the wine-growers thought they were ruined, today they know that the cause of their trouble is a miserable little insect and that a method of destroying it has been found.[101]

The next suggestion to come forward, mostly, one presumes, for the purpose of deflating the scientists, was that the attack was nothing new; it had happened before. The idea was comforting at any rate, for if the pest had been known before and had then disappeared of its own accord, it might well do so again. It is a common human trait. Ignore a trouble—the poor, today's hydrogen bomb—and it will go away. Among others, M. Koressios of Athens said it was the *phtheir* described by Strabo, and Herr Nourrigat, of Lunel, that it had laid waste European vineyards in the eighteenth century and was then called *gabel*.

Planchon became very indignant at these sorts of statements. He says:

> *A priori* the suggestions defy common sense. Can one imagine such an insect remaining quiescent for centuries of vine growing and then suddenly breaking out in such devastating quantities? Suppose even that it had to wait to do this for favourable weather conditions, why would it appear simultaneously at widely separated places, such as the French Midi, Bordeaux, Austria, Erfurt and Portugal?

As late as 1883 Prosper de Lafitte discusses the possibility that a 'worm'[107b] recorded as attacking vine roots in the twelfth century at Egaddi was the phylloxera, and considers it an unproved possibility. Yet another writer (Barral[18]) points out that there are still some growers who will not admit that the phylloxera is the cause of the trouble. They may be likened to today's flat-earthers. How we love to complicate life! The simple, proved, obvious proposition that the phylloxera was an indigenous American insect was too simple for a considerable number of people (*see* Note 6).

That it had occurred in Europe in the fifteenth century was also suggested by M. Fauveraux in his *Histoire de l'abbaye de Cîteaux*.[68] From 1420 to 1467, says this author, the monks of Cîteaux, proprietors of the Clos Vougeot vineyards, could not pay their annual rent of 13 *muids* of wine (?24 hl.) because their vines had died. First they perished in small amounts, but the deaths were soon spreading out from these places like an oil spot and the roots were covered with 'little beasts' by thousands. M. Fauveraux also maintained that the 'foreign' edition of Diderot's Encyclopaedia, under the entry 'insects', contained an article by Réaumur describing an insect which had destroyed the vines on the right bank of the Dordogne, from Bergerac to Ste Foy.

The standard French edition, 1751–65, of this great work contains neither anything like a description of the phylloxera nor an article by Réaumur, and it seems unlikely that this distinguished *savant* would write for one of the pirate editions, of which there were many. The trouble at Cîteaux certainly sounds like the phylloxera, but was very unlikely to have been; favourable conditions for root coccids and honey fungus are more likely causes for the monks' dilemma and, perhaps, M. Fauveraux's personal conviction of a European origin of the pest.

CHAPTER SEVEN

'The Butler did it', or, the Life History of the Phylloxera

The involved life history of the phylloxera aphid could be written up as a detective story, as bit by bit of its complex behaviour was uncovered. In this connection I am reminded of a drawing in the *New Yorker* of some years ago. It showed a strike picket parading outside a theatre in the typical New York style. Presumably a detective thriller was being shown. The men carried placards reading 'Don't patronise scabs' and so on, but the most effective anti-box office notice, carried by a woman with a particularly smug look on her face, read 'The Butler did it'. My readers already know the phylloxera did it, and the disadvantage of describing the insect's life history as it was pieced together is that of adding complications to complexity, so in this chapter it is set out as simply as we can make it.

When man, or, more likely, woman, began in the Neolithic age to interfere with nature on a scale hitherto unknown by inventing agriculture, he (or she) much affected the insect population and some of them became pests of crops. A little consideration will show that this was inevitable. Where a plant is growing in large numbers there will be found all those forms of life that can use it as food, including man, and which of them finally gets it is decided by a struggle between many forces. Darwin's great discovery was that those most fitted to survive in this struggle would, in fact, survive and thus give rise to the vast number of species on the earth.

There are some thirty Orders of insects and the phylloxera

belongs to the Hemiptera (the 'half-winged'; many insects in this order have half-horny, half-transparent wing cases). There are some 55,000 named species in this Order, many of them interfering seriously with our crops. All Hemiptera are sucking insects and they damage plants, whether directly by sucking the sap or indirectly by transmitting virus diseases. The Order includes plant-bugs, cicadas, scale insects and aphids, or greenfly.

Like the Holy Roman Empire, epitomized by Voltaire as being neither holy, nor Roman, nor an empire, the greenfly are not always green, nor are they flies, but are a big family, the Aphididae, to which the phylloxera belongs.

Greenfly, of course, are well known as pests; they frequently have complex life histories. Not only can they have different host plants for different periods of their life cycles, but most of them can reproduce themselves in two ways, by means of eggs, or by giving birth to living young (oviparously and viviparously). The eggs are usually the result of sexual mating and the living young are produced parthenogenetically, with no mating. The aphids in any given species can be winged (alate) or wingless (apterous). An example of a troublesome aphid with two hosts is the *Phorodon humuli* Schrank, the hop aphid, a considerable pest in most seasons, which divides its life between hop and damson.

The phylloxera has only one kind of host, the various species of Vitaceae, but a yet more complex life history. For instance, the species occurs as five different kinds of eggs and a dozen different forms of insects, of which only three are important from the point of view of damage done to wine vines (*see* Pl. 1). The life cycle is different according to the species of vine on which the insect is living. On the American vines the life is passed both above and below ground and sexual and non-sexual methods of reproduction are used. On the *vinifera* (European) vine the cycle takes place almost wholly below ground and only parthenogenetic reproduction is used.

Let us now examine the life cycle of an American vine. We start off with:

1. The winter egg, usually found in a crack in the bark of two-year-old wood. In spring this hatches to
2. A fundatrix nymph, females only being produced. The nymph

climbs up to the upper side of a young leaf, inserts its proboscis, and sucks in sap as food. At the same time she injects a substance into the leaf which causes the cells to develop in a certain way, in fact to form a gall, protected at the entrance with hairs (*see* Pls. 3 and 4).

3. The nymph becomes adult and lays
4. Yellow eggs in a circle inside the gall. These eggs hatch and become
5. Gall-living female nymphs, which creep out past the protecting hairs, settle on the same or another leaf and start producing their own galls. They grow into,
6. Gall-living adults who lay eggs in the gall in the same way as the fundatrix adult (No. 3 above). Three or four generations of gall-livers may be produced, becoming less fecund with each generation, but still alarmingly fertile. Whereas the fundatrix adult lays 500 to 600 eggs, the fourth generation female lays but 100. This means that one winter egg is potentially the forebear of 4,800 million gall-living females by midsummer, the time of the fourth generation. Among the third and fourth generation gall-living nymphs will be found some
7. Migrant nymphs (all female still). These move down to the roots of the plant where they settle, suck sap and become
8. Root-living adults. These settle either on (i) lateral parts of young roots, producing some deformities (phylloxera galls) and tending to prevent increase of root diameter, or (ii) on the point of a young root, near the cap-forming layer (the calyptrogon) where they stop its growth by the mechanical action of the mouth tube and the injection of saliva. The root-living adults lay
9. Root eggs which later hatch into
10. Root-living nymphs which become
8. Root-living adults which lay
9. Eggs.

   Three or four generations are passed in this way (stages 8, 9, 10). As the weather cools some eggs (9) hatch to either stage 11 or stage 12.

11. Sexuparous nymphs
12. Overwintering, or resting, root forms (rare on American

species). In the spring they continue the root-living cycle, 8, 9, 10. On some stage 11 nymphs wing buds develop and the nymphs become

13. Winged oviparous female adults which come out of the ground and find leaves either by climbing or flying short distances. These females conceive two eggs each, only one of which is laid, on the upper surface of a leaf. The eggs are either stage 14 or stage 15.
14. Eggs producing males or
15. Eggs producing females. The female-producing eggs are larger than the male-producing ones. In due course the eggs hatch, giving rise respectively to stages 16 and 17.
16. Male nymphs and
17. Female nymphs, growing to
18. Male adults and
19. Female adults. After mating the female descends to the two-year-old wood and lays a single
1. Winter egg. The cycle is complete. (*See* Fig. 3.)

This extraordinarily complex cycle is the usual one on American vine species. The survival value of an underground stage, and particularly of a winter egg, is very great for it facilitates survival through both hot, dry summers and severe winters.

On the European vine (*vinifera*) the American life-cycle can take place but it is most unusual. Both winter eggs (stage 1) and leaf galls (stages 4, 5 and 6) are rare and the life cycle consists almost entirely of the parthenogenetic root-living forms (stages 8, 9 and 10) repeated endlessly with the winter passed as the resting stage (No. 12). American vine species vary in their susceptibility to the pest and the longer the time the insect spends in the leaf gall form the less damaging is it to the plant, which is consistent with the *vinifera* being much more susceptible to the poison injected than are the American vines. On *vinifera* the galls and deformities produced on the roots (*see* Pl. 2) do not just stop growth or increase of diameter; instead they kill the roots. The insects, on finding their food source drying up, withdraw their sucking tubes from the roots and scramble out of the soil, by way of soil cracks and fissures, to the surface, where they walk about looking for a new vine. If they find one they descend to the roots and start feeding again. This is the main way by which the

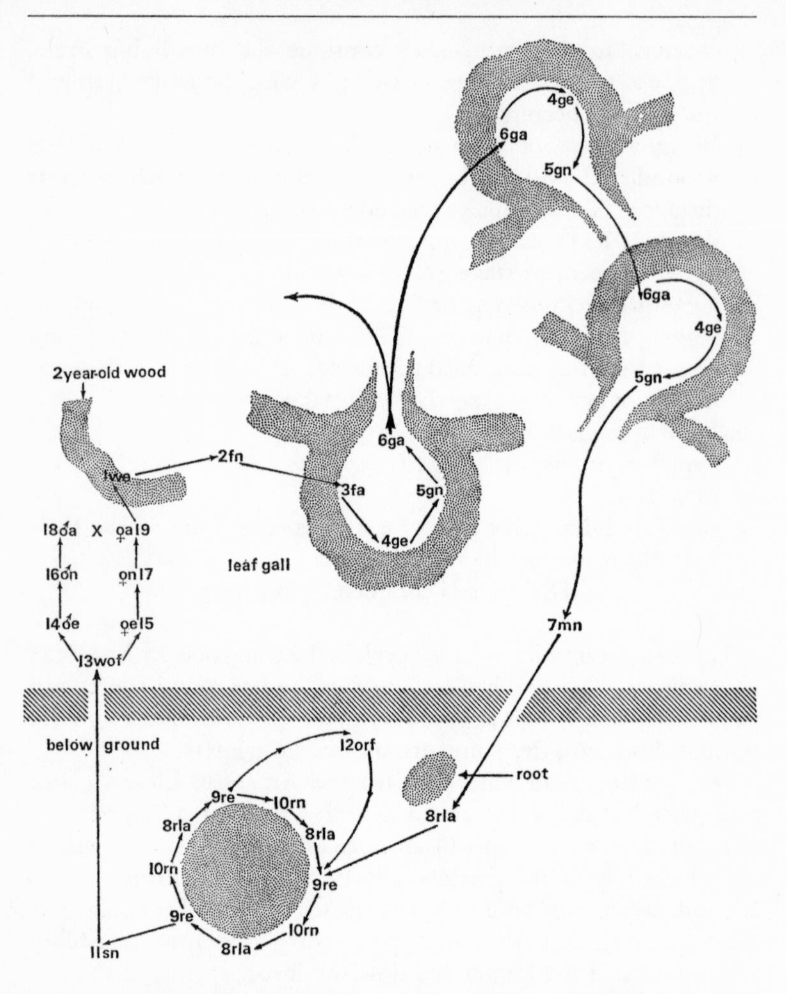

pest spreads, and it was this 'spreading oil spot' which gave the clue to Planchon and the early scientists in the late 1860s.

The complicated story was difficult to unravel for two reasons. Firstly, the idea that the life history could be different on the *vinifera* and the American vines had to be discovered and accepted, and secondly the much more difficult concept had to be swallowed, in that age of Victorian male superiority, that here was a species that throve without any sexual stage at all, that got on very well without the use of males; not the sort of news one would want to get into the hands of Georges Sand, Louise Michel and the pioneers

3. *The life-cycle of* Phylloxera vastatrix *on the American vines.*

1*we*	*winter egg*	*females only*
2*fn*	*fundatrix nymph*	
3*fa*	*fundatrix adult in gall*	
4*ge*	*gall egg*	
5*gn*	*gall nymph*	
6*ga*	*gall adult*	
7*mn*	*migrant nymph*	
8*rla*	*root-living adult*	
9*re*	*root eggs*	
10*rn*	*root-living nymphs*	
11*sn*	*sexuparous nymphs*	
12*orf*	*overwintering root forms*	
13*wof*	*winged oviparous females*	
14♂*e*	*eggs producing males*	
15♀*e*	*eggs producing females*	
16♂*n*	*male nymph*	
17♀*n*	*female nymph*	
18♂*a*	*male adult*	
19♀*a*	*female adult*	
1*we*	*winter egg, producing females only*	

of women's rights in France at that time. Possibly the basic secret fear of men to this day is that, biologically, the male is the less important sex.

The reproductive power of insects is enormous and only very few have to survive from each generation for the species to continue; with an insect feeding on a crop the survival rate has to be only slightly above normal for the insect to become a severe pest. This can be seen with phylloxera reproduction. One fundatrix has a potential progeny of 4,800 million above-ground insects in the season; as the insect weighs about 1 mg. this means that, if all survived, some 5 tons of insects would be produced from one stem mother. About five of them per hectare growing and surviving at this rate would give a bigger weight of insects than grapes! Of course they do not grow and survive at this rate; the wastage is enormous. Millions fall by the wayside and fail to find a vine leaf; birds and predacious insects eat millions more but, in spite of all, enough survive to spread the pest and maintain the species to this day. It is a wasteful process, but it succeeds.

We now need to know something of the appearance and behaviour of these twenty different forms of phylloxera and the galls they make, though we need not go into details of every one at great length. All these insects are female until stage 14, and obviously many of the features described can only be seen with a good hand lens or the low power of a microscope. The WINTER EGG is yellow when laid, but soon turns dark brown. It is about 0·28 mm. long by 0·12 mm. wide. Just before hatching in the spring it becomes amber coloured. It is 'sculptured' with a honeycomb pattern and has a little hooked stalk at one end which serves to attach it to the vine wood.

The FUNDATRIX NYMPH hatching from this egg is clear yellow in colour, about 0·25 mm. long at first and growing, after three months, to 0·45 mm. when adult. The nymphs coming from these eggs always move upwards to the leaves, generally to the last and youngest leaf on a shoot; at times they even start to feed on an opening bud. All attempts to get them to settle on roots fail. They establish themselves on the upper sides of the young leaves (often no more than 1 cm. in diameter) and start to feed. The feeding affects the leaf growth. As a phylloxera insect sucks the sap from the parenchyma tissues a small depression forms and the insect sinks into it. The depression swells and the insect continues to sink into the globule being formed. The edges of the gall thus shaping draw together, making a short slit in the upper surface of the leaf as entrance (or rather as exit); here a number of stiff hairs form, crossing each other in such a way that they prevent the access of parasites or predators of any size (for instance, they do not exclude predacious mites), but allow the young larvae to leave (*see* Pls. 3 and 4). Inside, the gall is smooth and round and the fundatrix female lives quietly there sucking sap and growing, and when adult, laying small, yellow eggs in what is both home and tomb, for she never leaves, this fact suggesting the generic name *Peritymbia* to Westwood.[186] Outside, on the lower side of the leaf, the gall is irregular, scabby and covered with coarse hairs, somewhat longer and more irregular than the usual leaf hairs. The gall itself is of about 3 mm. internal diameter, 4 or 5 mm. high and its walls are several millimetres thick; it forms from the growing tissue of the leaf. Considered objectively, it is a very remarkable adaptation of the leaf by the insect, for we have not only the protective cell but also the one-way exit, all induced by the sucking of sap and, presumably, injection of certain growth-regulating chemicals. One speculates as to how many centuries of development, of give and take, in the struggle between the vine and the aphid went by before the present situation was stabilized, and one wonders also if change and the struggle still continue.

Inside the gall the fundatrix lays a large number of eggs—which soon start to hatch. At times the new eggs are so numerous that the older eggs, hatching eggs and larvae are pushed out of the slit. Exhausted by egg-laying, the fundatrix dies and dries to a black shrivelled skin; then one or more larvae usually establish themselves in the gall, become adult and continue to suck sap and lay eggs. The gall also contains the skins cast by the fundatrix as she moults and, likewise, those of any of her descendants who establish themselves there. Each succeeding generation (always female) differs a little from its parent, becoming each time a little more like the root form until the difference between the leaf forms and the root forms is small, except in the antennae.

The FUNDATRIX is wingless and, at first, somewhat pear-shaped; she becomes rounded when full of eggs. The skin is smooth and free from tubercles, the main characteristic distinguishing her from the root-living forms. The mouth parts are long and fit into a groove on the ventral surface of the body. By its successive changes the insect appears to be getting ready to inhabit a new world, that of the soil. It is as though we were deliberately to breed up through successive generations a race of men better adapted to living weightless in space, as no doubt we shall before long. However, the aphids have the advantage over us in that a generation for them is a few weeks. Each generation becomes successively less fecund, a great biological advantage to the race, but a fact which led many scientists to draw a wrong conclusion—that the decline would continue and the species be extinguished in due course if sexual pairing could be prevented.

In autumn, when the leaves fall, all eggs, nymphs and adults in the galls die. However, before this happens and, in fact, throughout the second half of the summer, some of the larvae, by now beginning to resemble the root-living forms, move down to the roots, where they become root-living adults. The ROOT-FORM ADULT is a little smaller than the leaf-gall adult and has a number of brown tubercles or warts on its back. Its antennae are much notched and somewhat wedge-shaped. No more than a hundred eggs are laid. Like the leaf form it goes through three moults. The brown tubercles are arranged in lines across the breadth of the back and along the insect's length. There are twelve on the head, twelve on the prothorax, eight on the mesothorax, eight on the metathorax, six on the first ring of the abdomen and four on each of the remaining six abdominal rings—seventy warts in all. There are none on the posterior anal segment. Each tubercle has a short central hair.

It is at times necessary to distinguish between leaf and root forms, and it might well be thought that the presence or absence of these tubercles would readily identify them, but the insect moults three times, and after each moult the tubercles disappear. If you look at a gall-living form with a microscope you will see thickened skin and hairs at the spots where the root forms have hair and a tubercle. Thus a recently moulted root form can look like a gall-liver. The best distinction is that the leaf form will have no brown pigment at the tubercle sites. The hairs help the insect to get rid of its old skin at moulting.

FULLY WINGED SEXUPAROUS ADULTS have two pairs of transparent wings of a spread of about 3·04 mm., a body length of 1·14 mm. and a breadth of 0·38 mm. The body is greenish yellow, fusiform, with the abdomen tapering towards the anus. The head is broad and the eyes are large and red. The antennae are short, strongly ringed and without any marked tubercles. The wings are slightly opaque, white and delicate, and when at rest are carried pentwise (that is like a cottage roof) which is unusual in aphids. The three wing nerves are pale yellow and the wing can have rosy tints at times.

We must now return to the root forms and the autumnal change mentioned. As the weather cools a resting stage intervenes, and aphids in this stage are found in such places as under the bark of the thicker roots. Here the young root or gall forms take refuge from too great humidity and direct contact with the cold soil. They are usually found massed together in small groups, brown, flattened and roundish, looking somewhat like tiny tortoises. Their feeding tubes are inserted into the root tissues and they remain immobile the whole winter with antennae and legs drawn in. They are usually very small—first instar—not having reached the first moulting stage. Nevertheless a few adults may be found who slowly continue egg-laying, most of which fail to hatch. Great cold does not seem to kill the hibernating forms. For instance, in the severe winter of 1879–80 the air temperature around Orléans fell to −30° C. but the insects were not affected. Dr Horrath of Budapest exposed the overwintering forms to temperatures between −1° to −12° C. for eighteen days without affecting them. On the other hand, the soil temperature is important; at Montpellier it has been found that in spite of low winter air temperatures the soil at 25 cm. depth never drops below −1° C. In the Hérault overwintering forms renew their activity about the middle of April.

How much the insects affect the roots depends on the size of the roots and the species of vine. Large roots of any vine are not much affected, but in any case are not likely to be colonized by the insects. Roots of certain species, such as *riparia* and *æstivalis*, have reached a *concordat* with the pest and can put up with it, though they are not much attacked; small roots of *vinifera* and some American species (such as *labrusca*) are eventually killed by the insect.

Maxime Cornu spent five years studying the different galls and swellings caused by the aphids,[46 47] much of the money for it being subscribed by the *vignerons* of the Cognac. Cornu carefully followed the action of aphids on individual rootlets for some four months, making drawings of them from time to time, which in due course became a series of very fine hand-coloured engravings. Cornu found that when an aphid settled on a young root and commenced feeding, the root started to swell on the opposite side to the feeding puncture and to curve round the insect (*see* Plate 2), leaving it in a depression in which the eggs would eventually be laid. Two or more aphids are liable to settle near each other, in which case the shape and direction of the swelling will be determined by the respective positions of the feeding insects. Thus if two aphids are opposite each other the root will swell but continue to grow more or less straight. The swollen part of the root does not lose its power of producing adventitious rootlets, but when they are produced they in their turn are usually attacked too and further swellings produced. The attacked rootlets and swellings turn from white to golden-yellow, to brown, dark brown and black, at which last stage they are dead. There is usually a considerable starch accumulation in the dead root, thus making it more liable to attack by saprophytic fungi. It thus seems that the insect injects a growth-promoting substance which penetrates to the cells around the feeding puncture, leading to abnormal growth, but that the aphid then reabsorbs this substance from the cells immediately proximate to the puncture. Then the cells remote from the feeding point grow and those near it do not, thus producing the tubercle and the curve. Much the same effect was produced by the aphids on the roots of American vines, except that, on the resistant ones, the swellings lignified and the roots did not die.

Cornu [46 47] ends his paper with an account, and a fine plate, of other swellings and knots found on the vine and other plants, which are not the phylloxera but which, at times, have spread panic among farmers. These were the nodosities on the roots of legumes (beans, clovers, etc.) caused by the nitrogen-fixing bacteria, and similar swellings caused by eelworms, which last are now becoming a threat to vineyards. It must be remembered that this was ten years after the original discovery by Planchon, Bazille and Sahut.

Among a score or more of scientists in this highly technical field (elucidating the pest's life history) the principal workers were Signoret, Lichtenstein and Cornu; assistance was also given by Westwood in England, Riley in America and Balbiani in France. Signoret was first in the field, 1869,[164] absolutely accurate, though he drew the wrong conclusion (phylloxera/effect). Lichtenstein [114] was also accurate, but complicated his account by wanting to squeeze the facts into a new 'botanical' theory. Nevertheless he was a modest man and starts his article with a quotation from a play:

> 'Appoint me schoolmaster. . . .'
> 'But you can't read. . . .'
> 'That's how I'll learn.'

This was to make the point that he was an amateur entomologist and was writing for the practical man. He was really a botanist, and he thus tried to present the genus's complex life histories in a botanical form. His neatest outline of the new theory is in a letter he wrote to the great English aphidologist, G. B. Buckton, dated 10th September 1882; it is in very good English.[32] Lichtenstein points out that his theory is ten years old and is to the effect that the evolution of plant lice (aphids) is entirely different from the common metamorphoses of other insects. It is more like that of a plant, and Lichtenstein sees a budding process (*bourgement* or *gemmation* in French, *Keimung* in German) when the female lays eggs, or often but one egg, which produce only females. The theory was never accepted and there is no need further to explore its complexities. So strange was the life history that at times Lichtenstein was called the '*romancier du phylloxera*'.

Balbiani's early contribution [15 16] was important; not so much for his direct work on the pest as for his observations on other insects of the same genus, such as the oak phylloxera. Balbiani pointed out that a study of the other members of the genus would not necessarily throw light on the vine pest, but, nevertheless, such a study should not be neglected. In point of fact the life history of *P. quercus* of the oak was useful, for there were similarities to the life of *P. vastatrix* on the American vine. There was a winter egg, parthenogenetic reproduction and eventually production from eggs of sexuals, but there were no root forms. With *P. quercus* a winter egg was found in the bark crevices of *Quercus coccifera*, the Kermes oak. From this a female fundatrix arose and fed on the leaves of the same, evergreen, tree; her descendants grew wings and migrated to another species of tree, the sessile or durmast oak (*Q. pubescens*), where they fed from the undersides of the leaves. Sooner or later descendants of the durmast oak-feeders grew wings and flew back to the Kermes oak where they laid eggs, out of which male and female aphids hatched. These mated and the females laid eggs on the Kermes bark, thus completing the cycle.

C. V. Riley came to France in 1869 and identified the vine pest as being the same insect as that found in America. His first account [154] (1871) was in the third report on noxious insects in Missouri; it does not do much to uncover the insect's life history, but has a plate of seven different stages, probably redrawn from

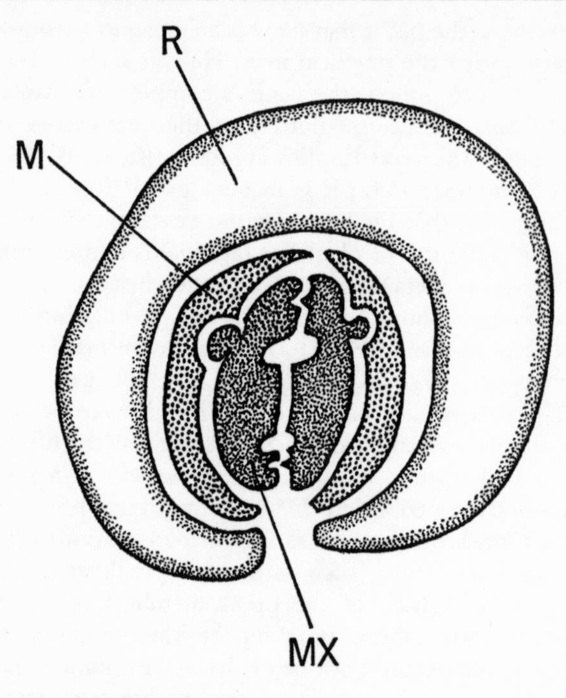

4. *Section of feeding tube of an aphid.*
R *rostrum*
M *mandibles*
MX *maxillae*

The parts press together to form two tubes. Plant sap on which the insect feeds rises in the upper one and saliva travels down the lower channel, having disastrous effects on vinifera *vine roots.*

French publications, such as Signoret's 1869 paper.[164] Riley's report is divided into two sections, practical and technical, the latter being concerned mostly with the insect's correct name and systematic position. It is really only in his report of 1874 that he goes into the natural history, and he completes this the following year, all the time giving very sound advice to the American grape growers.

Aphids feed on sap by inserting a fine tube into the plant tissue. In fact the sucking mouth parts of insects in general, and aphids in particular, are complicated and such complexity has an important survival value for the insect. The sucking tube itself is formed by two separate half sections, suitably grooved, being pressed together. It is as if a piece of thick double rubber pressure hose were cut longitudinally and then pressed together again to reform the twin tube. The two halves are the maxillae. They are surrounded by two further half-sections, the mandibles, and both these are adaptations of parts found in the mouths of biting insects (*see* Fig. 4). The two maxillae and the two surrounding mandibles are held together by muscular action of the insect and are assisted too by being placed in a groove of the rostrum, a more substantial organ protecting the mouth parts and often having a barbed point to assist in penetrating plant tissues.

In addition to the sucking tube there is also an ejection canal for saliva. As the sap comes down one channel the saliva goes up the other and assists the rostrum to penetrate the tissues; in the phylloxera the saliva carries the substances causing the gall to form, or the *vinifera* to die. In the mosquito it carries a chemical causing the blood to flow and, after an interval, an irritation (an interval is necessary in order to let the insect get away after feeding). The ejection canal can also carry substances or organisms causing diseases in plants and animals, such as virus troubles in potatoes or malaria (from *Anopheles* mosquitoes) in man.

The advantage of these complex mouth parts to the insect is that they can be cleared. If the tube gets blocked by the sucking of a solid particle the insect can withdraw its mouth parts, relax its muscles, separate the two halves of the tube and clean them. Otherwise a grain of dust might stick there and the insect would then starve to death. The madly complicated tubes and rostrum thus seem to be a piece of splendid Panglossion perfection.

It is now thought that the sap pressure within the plant forcing the sap up into the aphid is at least as important in feeding as is any action of the insect's 'pharyngeal pump'; this is why the insects abandon a plant so readily on the fall of sap pressure, and this early abandonment was a main factor in the failure to recognize the phylloxera as the cause of the 'new disease'.

CHAPTER EIGHT
Search for a Cure

The French national press pretty well ignored the new disease in vines. Far more exciting things were afoot. In the first place, in 1868, there was the Great Exhibition in Paris to describe, let alone the politics; and, secondly, farmers were always complaining about one thing or the other. These good peasants had but to work hard, be content with their lot and all would be well. Had not the oidium threatened to ruin the wine growers? And in a few years that threat was overcome. Space was too valuable to waste on such trivialities.

In 1869 Paris was interested only in business prosperity, having a great belief in the perfections of the liberal empire. Ollivier became Prime Minister of France, the alliance between France, Italy and Austria was established and the Suez Canal was opened. Obviously the world was changing very rapidly and, provided that 'dangerous radicals' could be kept at bay, prosperity for all, even the working classes, could not be stopped. Alas, in 1870 the acceptance of the Spanish throne by Leopold, Prince of Hohenzollern, offended Louis Napoleon, who declared war on Prussia under the mistaken impression that the other German states would support him rather than Bismarck. As we all know, Louis was soundly beaten. Meantime the phylloxera was steadily spreading. By the summer of 1872 *The Times* had noted it,[174a] repeating a report in the *Wine Trade Review* that the grape crop was but small, as good sunshine does not make up for grapes lost due to *coulure* (*see* Note 7), phylloxera and an insect known as the *ecrivain* (*Adoxus obscurus* L.). The Médoc would pick only half the quantity of grapes picked the previous year. In December *The Times*[174b] notes that a French scientist, M. Cornu, had been sent to Bordeaux to study the pest, which it misspelt 'philloxera' and misreported too, for the paper said it dried up both vines and fruit trees. (Of course it was not attacking the latter.) No remedy had been discovered but, says *The Times* somewhat prophetically, it must be traced back to America to find out what restrains it there. It is clear that they were thinking of biological control. In July 1873 *The Times*[174d] had an account of the 'new vine disease' in Portugal, which might have been either the phylloxera or the oidium; more likely it was the former, though it might have been both. M. Branas[29] estimates that the insect was in Portugal in 1863, though not positively identified until 1871. In January 1873 the French Government had introduced a Bill for the repression of drunkenness, which they thought would be best achieved by promoting the drinking of wine as a substitute for 'strong spirits'. As the phylloxera threat grew, so did fears that spirit drinking would increase. On 18th February 1873 there were, it was reported, twenty-three arrests in Paris for drunkenness; this does not really seem to be a very large figure for a big city, though it was a Tuesday. More cheerful news was that the government had 300 million francs in hand for the payment of the last instalment of the reparations to Germany and could thus secure the departure of the occupying troops. The rapidity with which the indemnity was raised is remarkable, and is a testimony to a touching belief both by the French and Germans, in fact by the whole world at that time, in the 'intrinsic' value of gold. A mixture of patriotism, love of a sound 6 per cent (*see* Note 8) and pride in the new democratic republic led the peasant and *petit bourgeois* to extract the *Louis d'or* from woollen stockings, mattresses and *petites caisses* to buy the reparation bonds in large amounts. An equally strange belief led the Germans to accept 1,460 tons of gold in full settlement, plus, of course, two provinces, not regained by France until 1918 (*see* Note 9). This is not the place to discuss whether the *Louis d'or* is merely a banknote printed on gold or whether it has a real value. The point is that great optimism reigned at that time (September 1873), and as far as the general public was concerned no nasty little insect was going to disturb it.

On 5th September 1873 the last instalment of the war debt to Germany was paid and the wine merchants of Bercy gave a great banquet the previous evening to celebrate the event, apparently quite unperturbed by the threat of a far more serious enemy than the Prussians. But by November *The Times*,[174e] always interested in the vintage, had a report from 'an occasional correspondent' at Boulogne (hardly a vintage area, it must be remarked) that the early promise of a good wine year had not been fulfilled; although the oidium had been overcome, because of the ravages of the phylloxera the crop would only be 37 million hl. It had averaged 50 million hl. In 1863 and 1869 yields were 70 million hl., but the current year (1873) at 37 million, would give a reasonable, even good, return at the price of 30–31 fr., that is, 1,100 million fr. or £43 million. Wine yielded the treasury 350 million fr. in taxes and the local authorities took another 75 to 80 millions in *octroi* levies. There was disappointment that a cure had not been found. 'Savants and empirics now vie with one another in monographs, suggestions, remedies and statistics; yet the cure is undiscovered.' The paper also discusses proposed remedies, for instance, the insecticide carbon bisulphide and flooding, the first of which it finds worse than the pest and the second, which would mean constructing a canal from the Rhône to the Hérault, Vaucluse, Gard and Drôme, is no good because excess of water is nearly fatal to the plant. M. Planchon had returned from his visit to the U.S.A. by this time and *The Times*'s Boulogne correspondent credits him with the view that the reason that the phylloxera was not a pest in the U.S.A. was an *acarus* that attacked the insect. Needless to say the *acarus* was not the reason, nor was it Planchon's view that it was.

Fortunately in the early days the scientists, local authorities and the government were not as uninterested as the press. As we have seen, in 1869 the Société des Agriculteurs de France appointed a Commission to examine the matter, the most important member being Planchon, though the others nearly all became phylloxera 'stars', to use the modern idiom; they were: the Vicomte de la Layère, M. Gaston Bazille, Dr F. Cazalis, the Comte de Lavergne, MM. Lichtenstein, Henri Marès, de Parseval, Planchon and Sahut, Baron Thénard and M. L. Vialla. The only stars' names missing are Maxime Cornu, Laliman, Balbiani and Signoret. The great interest and stimulus to investigations given by l'Hérault can be seen from the fact that four out of the Commission's eleven members were from this province.

Not to be outdone by a farmers' association the government next appointed a 'Superior' Commission for the phylloxera; it was originally set up to adjudicate on the award of a prize for a remedy. As the problem grew and still remained unsolved so did the 'Superior' Commission grow, and the law of 15th July 1878, modified by decrees of 6th September 1878, 27th February 1879 and 5th July 1879, made it into a powerful and knowledgeable body.

As well as the attention given to the new pest in the provincial and national presses there was considerable interest shown in British diplomatic circles. The Consul at Oporto, Mr O. Crawford, reported, in a letter dated 3rd July 1892,[78] that the new vine disease at Oporto had created fears almost amounting to panic in some districts, which he attributes to a pamphlet compiled by a Senhor Oliveira, 'a hasty and ill-judged compilation from the French authorities'. This shows how difficult it is for a layman to judge a technical problem. Consul Crawford concluded that no serious damage had been done, that the majority of vineyards in Portugal would escape and that the phylloxera, if it attacked at all, would do so first in the Bairrada district and then in the port wine area. One wonders what the basis of Mr Crawford's opinion was, apart from having in mind King Henry IV's 'Thy wish, Harry, was father to the thought'. His superior officer, Sir Charles Murray, the British Minister at Lisbon, was not so sanguine as the consul. Sir Charles had written to Earl Granville on 12th July 1872 in a quite different tone: '. . . the new scourge is as devastating in its effects as the disease called the oidium . . . terrible in Douro . . . a yield of 1 pipe in a particular vineyard instead of 70. A commission had been appointed.' Sir Charles then gives some details of a French remedy, to dig a space around the vine and fill it with soot. Ever with an eye to a market, he concludes: 'If this should ultimately prove an effectual antidote to the malady, it is much to be regretted that the vine-growers of France and Portugal are not somewhat nearer to London where it [soot] could be cheaply and abundantly provided.' The soot may well have contained some nitrogen and have helped a temporary recovery of the vine.

The British Consul at Bordeaux was at first not unduly alarmed; on 9th July 1873 he sent the Foreign Office a copy of a communication from a M. Duval to a M. Dupont saying that the phylloxera was found throughout the Gironde and that such was the general state of alarm that all vine deaths were now attributed to it. 'This is a serious and profound error.' M. Dupont was general secretary of the Société d'Agriculture and he was not as convinced of the matter as M. Duval. The British Consul sent details of the prize of 20,000 fr. (£800) to be awarded to the inventor of a cure. The idea had to be sent to the Minister of Commerce with an assurance that it had been tested and found successful. The suggestion would be then sent to the central Commission, who would try it out and indicate the winner, if any, this last provision being the sting in the tail. The closing date was 31st December 1872. This gave any serious competitor very little time to carry out his tests, and in contrast it gave every opportunity to the lunatic fringe and charlatan.

The value of the Hooker/Planchon contacts may here be seen. Most of these consular reports, together with the newspaper cuttings and reports of the bodies concerned with phylloxera, went to Dr Hooker at Kew for assessment and he, having been in constant correspondence with Planchon and thus in a better position to judge the facts reported, wrote to Viscount Enfield on 13th December 1872 [106a] that the phylloxera was a serious pest and that every effort should be made to keep it out of the British vine-growing colonies.

By 18th June 1873 the French Government was sufficiently alarmed to offer a large prize (300,000 fr.) for a remedy (Law of 22nd July 1874). In order to test the suggested remedies the government arranged with the Hérault Commission for trials to be made at the School of Agriculture at Montpellier, where an attacked vineyard was set aside for the purpose. It was known as 'Las Sorres'. In 1877 an interesting report [42] was issued on the experiments carried out.

Experiments with vines in general, and with phylloxera in particular, have a number of built-in difficulties with which we must deal before turning to the Las Sorres results. In the first place it is very difficult to say what a 'normal' crop of grapes is; the weight picked and the quality are so dependent on weather that both can vary enormously from year to year. Moreover it may only be possible to assess quality some *years* after the wine has been made. As regards quantity, fruit crops tend towards 'biennial bearing', good and bad years succeeding each other. Grapes do not regularly follow this 'on/off' pattern because they are so much more influenced by weather and are, in consequence, more unpredictable, although their coefficient of variation is less than that of both apples and wheat.

As a subject of field experiment the phylloxera also presents certain difficulties. In the first place it is on the roots and cannot be seen, and you cannot constantly be digging up a plant to see how it is getting on. The object of the Las Sorres tests was to see if an attacked plant could be cured by any of the numerous treatments proposed, but in any vineyard selected the plants submitted to any particular treatment might be on their last legs, lightly attacked or even healthy, the only clue to their condition being the general appearance of the foliage. It could thus be very difficult to know if the treatment had harmed the plant, for either the pest or the treatment might have done it.

The trials were carried out by MM. Durand and Jeannenot, two professors at the School of Agriculture, Montpellier, and secretaries of the Hérault Phylloxera Commission. They must have worked exceptionally hard to produce order out of the flood of suggestions reaching them, and they obviously were men of immense integrity, fully in the spirit of the University of Montpellier's tradition, dating at least from the days of Rabelais. A first series of phylloxera tests had already started on 8th May 1872 and ended on 8th June; they tried forty-six procedures, using 215 vines and had no success. They were using about five vines per trial and realized that this was too small a number; in later experiments they increased the plot size to twenty-five plants. The plots were thus five by five vines, marked out with posts and having a central post with a label bearing the name of the inventor of the process. Two rows of vines, each 1·5 m. from the next, separated the plots and also served as control, untreated plants.

Results were assessed by MM. Durand, Jeannenot and Sahut by giving marks for the state of vegetation, length of shoots, weight of fruit and sugar content of the must in the treated plots and the controls. The vegetation was an early indication of the health of the plant and two apparently different observations were made on it, its appearance * and state,† a

* *Aspect du Feuillage.* † *État du Végétation.*

distinction we have not yet quite appreciated, possibly how the vegetation looked (healthy or otherwise) and its abundance. A scale of 12 marks was given to each plot and the differences between the marks for treated and untreated plots entered in the final column of the result tables. It can at once be seen that the recording work was enormous. In each plot at least eight observations had to be made and the shoots on twenty-five plants measured. The Commission tested 317 processes, some of them several times, so that they dealt with some 16,000 plants. They had no adding machines or computers to deal with the vast mass of figures pouring out of the plots. Moreover their patience had to be enormous, for they were assailed by cranks and charlatans and entered into a vast correspondence with all and sundry. Moreover it all seems to have been done on the modest budget of 5,000 fr. per annum.

Table 3, pages 208–9, shows how the figures finally were presented, and Table 4, page 210, gives the sites and the number of trials made over the five-year period.

Up to October 1876 the Commission received 696 suggestions for remedies mostly sent by the Ministry of Agriculture, which may well have stopped some of the patently absurd ones reaching Montpellier. The two tireless professors did not follow up 379 proposals for several reasons—because they were duplicates of previous suggestions, because the authors would not disclose the active ingredients, because they imposed conditions the Commission could not accept, such as money for research and travelling, because the samples or proposals arrived too late or because they were just stupid, at any rate at that time. Unfortunately one knows very little about the rejected 379; no doubt many of the proposals were ridiculous, yet yesterday's absurdities tend to become today's dogma. Some years earlier the painting exhibitions had ignored or were hostile to the work of such artists as Cézanne and Manet, who were forced to exhibit in a *Salon des Réfusés*. Some of the *réfusés* of Montpellier are possibly methods in use today, such as the suggestion that 'the sap of the vine should be made poisonous to the aphid'. This sounded pretty absurd in 1874, but today we have systemic insecticides which are absorbed by the plant, make the sap poisonous to insects and do not harm it. Even in those days Planchon noted that carbolic acid appeared to be absorbed by the roots and transmitted to the leaves without damaging the vines; he traced it by the smell so that most likely it was some impurity in the crude carbolic fluid (rather than the phenol itself) which moved in the plant. Unfortunately it did not kill the insect (*see* Note 10).

It is strange to reflect that about this time DDT was first made by Dr Zeidler in Germany.[188] To him it was just a new chemical, and its remarkable insecticidal properties were not discovered until 1940. Had this compound managed to find its way into the Las Sorres tests a very different story would have been told.

Some clues to the Montpellier *réfusés* are Valery Mayet's remarks in 1890;[124] he comments on the flood of documents (it is yet greater today), although the matter was but twenty years old. 'What floods of ink and ineptitudes, what madnesses were put forward for the attainment of the 300,000 fr. prize!' and he quotes Planchon:

> To get to the bottom of this matter one would have to plumb floods of ignorance, to consider the idea of a living toad buried under the vine to draw the poison to it, to adjudicate on the proposal to water the sick vines with white wine or an infusion containing mallow as its main constituent. Among the deluge of suggestions most come from people confusing phylloxera with oidium, who have never seen either of them. The growth of this file of stupidity is a sad reflection on public instruction in scientific matters. Wild theories reach us from all ranks in society and from all parts of Europe. The mostly highly recommended to the Minister of Agriculture are usually the most stupid; the most tenacious are those bright spirits who ride a hobby-horse to the borders of madness. Fortunately, as we gain experience, we can see the problem more clearly and push these follies to one side.

The toad suggestion seems to have come from a classical scholar, for it is a remedy against storms and pests found in Pliny's *Natural History*,[147] but Planchon does not seem to have recognized it. We have already mentioned his remarks on the failures of French education in this respect. Foëx too felt badly about some of the silly methods suggested.[75]

> None of the methods proposed for fighting the phylloxera has produced a greater number of inventions, illusions and deceptions, it must be said, than the use of insecticides. This field has given rise

> to the widest range of proposals, some at times, very strange: everything has been suggested, from toad venom to tobacco juice, not forgetting insect repellants, such as exorcisms or beating mechanisms by means of which the insect would be driven methodically into the sea or over the frontier.

In spite of the tremendous effort put into the Las Sorres experiments the results were almost entirely negative in that no process tried actually extinguished the phylloxera. In 1876 the two by now no doubt somewhat weary professors were able to accumulate a table of only thirty-two treatments where the difference column was positive, that is, where the treated vines were better than the control vines, and in only two of these did the treated plots show any marked advantage over the control plots, that is, where the treatment scored 6 or more (out of 12) over the control. These two treatments were potassium sulphide in human urine and another sulphide with colza cake. The sulphides tended to release the insecticide sulphur dioxide in the soil and the urine and cake supplied nitrogen to enable any few roots left unharmed to make a quick recovery.

A wide range of substances was not only proposed, as Foëx pointed out, but was also tested almost entirely on empirical lines, much as insecticides are tested today, the only difference being that we have a much greater range of chemicals to draw on. The index of substances in the Las Sorres trials gives some 224 different products used, but as they were used at different strengths and in different ways there are over 500 methods on which we can draw. There are over 750 names of people communicating with the school on the subject.

The substances tested fall into a number of classes. There is the usual range of inorganic chemicals, sulphates and sulphides of various metals, mineral and tar oils, odiferous substances (garlic, turpentine), drugs (quinine, rhue), vegetable insecticides (pyrethrum, tobacco), soot, ashes, acids, fertilizers (urine and cakes such as castor and colza) and a lunatic fringe which somehow got by (sealing wax; *see* Table 3, pages 208–9). There are also four 'secret remedies', which is surprising as the Commission's policy was not to test anything whose inventor would not reveal the active ingredient. The sealing-wax suggestion was put up by M. Louis-Phillippe of Annecy, Haute-Savoie: even had he noble connections one would not have expected them to carry much weight in the young third republic. As to the secret remedies, they were put forward by MM. Allier, Charmet, Flahaut and Fremont-Laffargue. Perhaps they were supported by such illustrious names that they could not be refused, for these gentlemen do not seem to have been important in their own right. Although a Flahaut was *aide-de-camp* to Napoleon III, that also was hardly a recommendation in 1873. Only one of the secret remedies showed any advantage over the control plots, and even that was but a one-point rise out of a possible twelve.

The 750 names in the index are also interesting, not least because of the absences; for instance, we do not find Prosper de Lafitte, Laliman (who later claimed the prize), Baron Thénard, the Duchesse de Fitz-James and many other people with reasonable suggestions for phylloxera control. Admittedly there is a number of anonymous entries in the list (there are five Monsieur X, for instance), and these may well conceal the identity of such people. However, only one of the MM. X. appears to have had his proposal tried—bones dissolved in sulphuric acid; it proved of no use. Naturally most of the entries are French, but people in a number of foreign countries also corresponded on the subject. One would expect to find those from the wine-growing nations such as Algeria, Austria, Italy, Spain and Switzerland, but nationals of other countries are there as well. We find names entered from Bavaria, Belgium, Brazil, Denmark, England, Holland, Java, Prussia (Berlin), Singapore, South Africa (Cape Town), Syria (Beyrouth) and U.S.A. (Boston and New York). The only foreign entries from outside wine areas that were considered worthy of testing were those of Mr Robby of Denmark (paraffin oil and water), Mr Cope of Birmingham, England (a decoction of elder leaves) and Mr Russell, of Albany, New York State, U.S.A. (drilling a hole in the vine stem and filling it with sulphur), and again none of them showed a positive gain over their control plots.

It was sad for MM. Durand and Jeannenot that this vast labour uncovered no great remedy. What did emerge was that insecticides, upon which such hopes had been built, were expensive and were but palliatives, and two schools of thought began to emerge, the chemical and the biological (or the 'Americanists' as they came to be called). The 1876 summary of the station's researches starts by praising the painstaking work of the two principals. In five years, though they had not found out how to destroy the phylloxera, they had accumulated a great deal of information which represented progress towards a final solution of the problem. Up to the present all attempts to kill the winter egg had been as useless as those made with insecticides to destroy the root forms. It had to be borne in mind that neither of these desired results would be secured by any of the methods tried up to the present (it will be noted that the Commission still held to the belief that if the winter egg could be destroyed the cycle would eventually be broken; it was quite erroneous). This meant that another method must be found; one would be somehow to make the vine resist the attack of the insect and this meant turning to American vines, from whose resistance much was to be hoped and expected. 'Do these experiments present us with any hope of giving the European vine any useful resistance to the phylloxera?' M. Marès asked, whistling, one feels, in order to keep up his courage. He thinks they do, but also pins his hope on some insecticides which, whilst not totally destroying the pest, will confer sufficient resistance on the plant as to enable it to crop more or less normally. The insecticide treatments favoured were soft soap, followed by farmyard manure, and certain potash salts and sulphocarbonates. M. Marès would like to go on with the work, particularly the sulphocarbonate work, grafting American vines and packing the soil tightly round the European roots, and he hopes government will continue their annual grant of 5,000 fr. again in 1877. This seems a very modest sum (£200) for such an elaborate experiment; presumably it did not include the professors' salaries. There was one difficulty though; the tests could not be done at Montpellier, for there nearly all the vines were already dead. One assumes that by 1876 M. Henry Marès had got over his original belief that the phylloxera was the *effect* of the trouble and not its cause.

There is one unusual thing about the insecticide treatments reported, the failure of carbon bisulphide; used alone or with paraffin oil in four different tests it left the treated vines worse than the controls, which is surprising because this chemical did kill the pest and was much used at one time. The rates used in the tests ran from 76 to 444 kg. per hectare, which was not excessive. It seems to illustrate the general difficulty of doing experiments with this insect. If its distribution were irregular, and chance saw to it that the control vines were less attacked than the treated vines, such a result might will be produced. Modern experiments of this nature use a multiple plot system, with a scatter of treatments over the whole area; a mathematical analysis of the results is then made, thus estimating what the chances are of any given result being due to chance or to some real difference between the treatments given.

The success or failure of a treatment in these tests, as we have noted, was judged mainly on the appearance of the treated and control plots. The authors point out, however, that this is not the only factor to be considered and that differences in the weight and quality of fruit picked are also important. The sulphocarbonate plots often gave two to four times more grapes than the control, untreated plots (*see* Note 11), a considerable encouragement to the earnest workers in this field. But it was not the result for which they were looking. They wanted to abolish the phylloxera for all time, and they turned to the possibilities of American vines.

In the spring of 1876 thirty-five kinds of American vines were planted along with French varieties in what is rather charmingly called an *école comparée*; they belonged to the species *labrusca*, *cordifolia* and *æstivalis*, and were interplanted with the French kinds (the *viniferas*). For some strange reason the researchers also grafted twenty-one American varieties on to still vigorous *vinifera* roots such as Aramon. The reason is difficult to find, because it was the *vinifera* roots that were dying; perhaps they thought the American sap would confer some immunity on the European root and that the European root would modify the terrible foxy taste of the American scion, a latent Lysenkoism. Perhaps they were so immersed in empiricism that they would have tried anything. The system was later used by the Duchesse de Fitz-James for a different end (*see* page 116), which we shall discuss later. Naturally most of the American varieties grafted on to the *vinifera* Aramon failed because of phylloxera attack on the roots. The Las Sorres soil contained about 30 per cent of chalk and at that time (1876) it was not known that American vines do not take kindly to this substance.

Another task undertaken by the Commission early in its life (1873) was the study of flooding. Watering the vines to help them recover had been found to be of very little use, but a M. L. Faucon, at Graveson, had gone far beyond this and flooded his vineyard for weeks on end; it appeared to keep the pest at bay. This suggestion was opposed by a M. A. Pieyre near by, who said he found that vines in the lower parts of his vineyard which were immersed for weeks at a time became stunted and useless. M. Pieyre had a rival cure, a handful of sulphur around each stem and a forkful (3 kilos) of farmyard manure. The Commission decided to visit both these gentlemen, and eleven commissioners, deputies, the inspectors of agriculture and so on, set out on 28th August 1873 and were joined by other officials *en route*. They seem to have been pretty thorough in their methods, questioning, tearing up vines and arguing. At Maillau, M. Pieyre's vineyard, the plants were reasonably healthy, but the pest could still be found, even on roots which actually grew through a patch of sulphur, and although a considerable improvement had been obtained over the worst year's crop (1870) of 100 hl. of wine (it was 230 hl. in 1873) it came nowhere near the pre-phylloxera figure of 1867—1,200 hl. At Maillau the stagnant water of winter did seem bad for the vines, but they were generally not well kept—couch grass was a bad weed throughout. The system did not seem very promising.

M. Faucon was obviously a very different man. In the first place he had neatly assembled all the facts, sat his audience down and read them a report on the matter, thus saving a great deal of question and answer and time. (It is a mistake to think that in the 'good old days' things were more leisurely and there was time for everything. There was much less time—travelling, for instance, took so long and restricted observation and debate.) M. Faucon's vineyard, Mas de Fabre, was on a deep clay soil (down to 10 m. in depth), strong and compact. The vines, five to eleven years old, were at 1 m. in rows 2 m. apart (5,000 per hectare), and the crops, areas and treatments had been as set out in Table 5, page 211.

Yields of 60 hl. per hectare before the attack had fallen to under 3 hl.; flooding had restored the wine made to 40 hl. per hectare, and this in a season that had suffered from a spring frost.

M. Faucon could get 60 l. of water per second from the Durance canal, enabling him to flood 3 ha. to a depth of 10 cm. in twenty-four hours. A fifth of the flow, that is 12 l. per second, served as a feed and kept the area flooded. It is obvious that a large volume of water was needed and some consideration should be given to the theoretical amount required and to loss of water from drainage and evaporation. In the first place 10 cm. of water above the surface of the soil over 3 ha. needs 30,000 × 0·1 m. = 3,000 cubic metres of water. A cubic metre of water is equal to 1,000 l., hence a 10-cm. layer of water will need 3 million l. of water standing on the 3 ha. The soil must also be saturated to a depth of, say, 1 m., and we can take the air space in a winter soil to be 10 per cent of its volume; 10 per cent of a metre is 10 cm., and the water needed is thus equal to that needed to give the supernatant 10 cm. In other words 6 million l. of water are needed to soak 3 ha. and get a head of 10 cm. over it. M. Faucon's delivery was 60 l. a second, consequently he could get 6 million l. in:

$$\frac{6{,}000{,}000}{60} \text{ seconds} = 27 \text{ hours } 47 \text{ minutes}$$

This is a little close if he were covering 3 ha. in twenty-four hours, and allows nothing for evaporation and drainage. Perhaps M. Faucon's flow was faster than he thought, or his winter soil was very wet and needed very little water to fill it. Moreover he says he needs a fifth of the flow to maintain the head, so presumably evaporation and drainage were claiming 12 l. a second, which makes the calculation fall still shorter of the theoretical. If the air space were only 5 per cent of the soil the need would be 4·5 million l., delivered in just under twenty-one hours leaving some three hours for the draining and evaporation supply, and perhaps this assumption is nearer the facts.

The Commissioners dug up vines and found no phylloxera, though in neighbouring fields they found patches of live insects driven out of the soil by recent rains. They thought M. Faucon's results very encouraging; though it had not re-established a completely normal root-system yet it could be used to restore vines 'devastated by the disease and the phylloxera'; evidently, in 1873, M. Henry Marès had not yet abandoned the phylloxera/effect belief. However, they did not point out that it ran contrary to experience, both in general and in the case of M. Pieyre, that vines covered with stagnant water perished: there was also the question of cost, which they set aside for the moment. Flooded vineyards would have to be fertilized to make up for the plant food washed out by the drainage water; it is possible that the muddy Durance water was supplying some fertilizing elements. It will also be noted that whilst Faucon had improved on anybody else he had only restored the yield to 60 per cent of the 1867 figure.

M. Marès ended his report by saying that he hoped more people would imitate M. Faucon and that the government would vote money for flooding tests to be carried out at Montpellier.

In 1874 M. Faucon also wrote a very persuasive book [67] on the subject of vine submersion, refuting his attackers, giving a great deal of practical advice and an estimate of the cost of flooding. For instance, he says though the system cannot be applied to all vineyards, this is no reason why it should not be used in those places which are suitable, more numerous than one might believe if a little thought is applied to the matter. In a few years' time the floodable vineyards will be the only ones left because the insect will have destroyed all others. His success is *not* due to any special soil or situation, or to the silt in the Durance water, which did not amount to much in any case, 0·1 mm. per year,* nor does he need to apply undue amounts of fertilizer. It *is* due to careful work and complete flooding to a depth of 10 cm. for at least forty days in autumn or winter.

In his practical section Faucon points out that the flooding must be complete; an island of even a single vine left above the water at once becomes a reinfection centre. The land must be prepared for the flooding by making flood basins and any earth banks used to make these basins must not have vines on or near them because the vine roots will grow into such parts and become a refuge for the pest. The remedy then? Ruthlessly to sacrifice the vines on or near such banks. The water used should be as airless as possible, thus if pumped it should not splash about or fall in cascades to entrap air. He also noted that the water seemed to follow the vine roots downwards and, though roots might go very deep, the head of water on an area would ensure penetration along and around such roots and kill all insects found there provided it remained long enough.

* Even so, this is equal to 1 cubit metre, say 1·5 tonnes of silt per hectare per year.

M. Faucon gives the following costs:

For 21 ha.

Preliminary work. Construction of water take-off, channels, furrows, sluices and water basin banks 3,000 fr.

Interest on above at 5%	150·00 fr.
Water fee, 21 ha. at 35 fr.	735·00 fr.
1 man to control operation, 45 days at 3·50 fr.	157·50 fr.
1 boy to help, 45 days at 2·00 fr.	90·00 fr.
Summer irrigation, 15 days at 3·50 fr.	52·50 fr.
Repair of channels, 15 days at 3·00 fr.	45·00 fr.
Miscellaneous expenses	30·00 fr.
	1,260·00 fr.

or 60 fr. per hectare, much cheaper than insecticide application. M. Faucon maintained that the figures were on the high side. He no longer needed a boy to help as the channels now needed scarcely any attention, having solidified through use.

That same year (1874) Aristide Dumont [7] proposed that a canal be constructed from the River Rhône at Condrieux (about 30 km. south of Lyon) to feed the vine areas. An inquiry was put in hand and the cost estimated at 80 million fr. The departments of the Drôme, Vaucluse, Gard and Hérault were greatly in favour of it and offered to raise 30 million fr. But there was also considerable opposition to the proposal, perhaps expressed most forcefully by the Duchesse de Fitz-James who characterized it as 'this watercress culture system'. The vine, after all, was obviously a dry land plant and flooding it for weeks at a time each year seemed destined either eventually to destroy it or to produce thin watery wine.

The success of the Faucon system can be seen by the fact that it is still in use. An example is the 4,000 ha. in the Ste Marie and Aiguesmortes area of vines on their own roots flooded every winter and giving good crops. The method is also used in other parts.

It must be remembered that in the mid 1870s the complete extinction of the phylloxera was being sought and provided this could be done costs were not of great importance. The fact that it could not be done with insecticides was a lesson learned painfully and slowly.

In 1872 Baron Thénard started to use the chemical carbon bisulphide as a control for the pest. We shall deal with this in the next chapter.

CHAPTER NINE

Carbon Bisulphide

Carbon bisulphide was discovered by the German chemist Lampadius in 1796, when he was twenty-four. Lampadius later distinguished himself by his work on metals, electricity and solvents for oils. Being easy to manufacture and a powerful solvent for these, carbon bisulphide was soon in use commercially.[26]

Baron Paul Thénard, himself the son of a famous chemist, became a member of the Académie des Sciences in 1864 and took a great interest in phylloxera control on the chemical side. He was held as a hostage by the Germans after their victory in 1870 and released after payment of the war indemnity. He owned a vineyard and, interested in the activity of volatile organic compounds, he conceived the idea of using carbon bisulphide (CS_2) for phylloxera control. The product is an effective fumigant at relatively high doses and is dangerous both from the toxic nature of its vapour and its liability to catch fire or explode. It is a heavy liquid (specific gravity 1·27) with an unpleasant smell, most of which is due to impurities. It freezes at about −112° C. Igniting spontaneously at 140° C., burning with a bright blue flame, it generates considerable heat, as might be expected since both of its component elements burn. A mixture of the vapour with air is easily ignited by a spark, a static one from some comparatively simple discharge or one from an electric motor or switch, or a frictional spark from, say, a flint in the soil. It will catch fire if dropped into a copper vessel containing boiling water. Carbon bisulphide inhalation is dangerous to man.[30] It was said at that time to be safe to him at concentrations below 10 to 3·3 parts per million. Symptoms are likely to appear after a daily exposure to 33 ppm or more, or after a single exposure to 330 ppm for a few hours. The chemical can be absorbed through the skin. An additional hazard is that exposure to low doses leads to an inability to detect the odour, so that a person may continue to work in a toxic atmosphere without being aware of it, thus putting himself at serious risk. Carbon bisulphide is liposoluble and a nerve poison. The first symptoms of poisoning are headache, irritability, wavy and double vision and mental confusion, leading to cramps, stomach pains and nightmares. Prolonged exposure to concentrations over 1,000 ppm for thirty minutes or more leads to excitement, coma, convulsions and death by failure of the respiratory centres.

In France such large quantities of this chemical were being used in the early 1860s for vulcanizing rubber that inquiries were made into its toxicity, the most notable being that of Professor Delpech,[55] of the Necker Hospital, Paris (*see* Note 12).

Carbon bisulphide was made by passing sulphur vapour over red hot charcoal. Cooled with water, the gas condensed into liquid carbon bisulphide. The product was led to zinc vessels and stored under a layer of water. If required for vulcanizing it was redistilled, but for agricultural use this was not necessary. The layer of water over this volatile chemical was a great safeguard, but a purchaser would need to know how much water was in his barrel. A simple test was devised. A stick or rod was coated with grease or tallow, plunged into the cask and withdrawn. The chemical dissolved the tallow, thus indicating depth of solvent in the barrel. The manufacturing process sounds frightening considering the explosive nature of this product and the possibility of leakage in the plant. A factory did catch fire and blow up in Libourne (Gironde) on 20th November 1873.

The insecticide properties of carbon bisulphide appear to have been revealed in 1854 by M. Garreau,[80] who made experiments in the killing of grain weevils in cereal stocks, using sulphurous acid, carbon dioxide, turpentine, camphor, chloroform, naphthalene and carbon bisulphide, this last being the only substance having any effect at strengths at all practicable. In 1857 a M. L. Doyère made similar tests in Algeria, with great success. The subsequent ventilation in the learned journals of a dispute as to who was the original discoverer drew general attention to the insecticidal properties of carbon bisulphide vapour.

Thénard's first experiments were done in 1869 on two estates in the Bordeaux area and were not successful. He used too much of the chemical and injected it too near to the vines, so that they were killed. His dosage was some 800 kilos per hectare. The experiments got tremendous publicity and appeared to be considered a major disaster. Actually he used the chemical on only eleven vines.[173] On M. Chaigneau's estate both the insect and the vines were destroyed and on M. Cahussac's the insects were not killed. The results were hardly catastrophic, and as it happened most of the 'dead' vines sprouted again the following August. Perhaps disappointment led to exaggeration. However, by 1873 no miracle cure had been found, and thoughts turned back to carbon bisulphide; M. Monestier, at Montpellier, tried much smaller doses with very good results. Much depended on the nature of the soil. Carbon bisulphide gave poor results on heavy clay soils and on the dry, shallow soils of hillsides. Soil moisture was also important; the earth had to be just right, not too dry (when the vapour would quickly escape), or too wet (when it would not penetrate to all parts and kill the insects). It proved very successful on light sandy soils, but here the nature of the soil itself was a help.

At this point it is of interest to consider how much carbon bisulphide was theoretically necessary to destroy the phylloxera in a given area of land. Comparatively recent experiments with carbon bisulphide show that 160 mg. of the chemical per litre of space kills the Japanese beetle [74] and that 65 mg. per litre kills the grain weevil.[128] Both these are pretty tough bettles and we are safe in assuming that 65 mg. per litre would kill the small soft phylloxera aphid. The quantity is equal to 65 g. per cubic metre (m^3). Vine roots we can take as penetrating to a depth of 2m., and assume that the winter air-space in the soil is 10 per cent of the volume. Consequently the volume to be fumigated per hectare is $\frac{2 \times 10{,}000 \times 10}{100}$ cubic metres $= 2{,}000\ m^3$.

At a rate of 65 g. per m^3 this gives 130 k. per hectare, the sort of figure eventually used; it will be noted that Thénard's first trials were at six times this amount. It must be remembered that there are two important variables in the calculation, the percentage airspace and the depth of the soil. A dry soil in summer may have 50 per cent of airspace in it; a wet soil none. The soil may also be a few centimetres deep (usually the best wines are produced on such soils) or of several metres, consequently dosages of carbon bisulphide have to be varied greatly to suit circumstances; the first users did not realize that this was a volume calculation, tending then to regard it as an area one and talking about quantity per vine or per hectare.

A number of 'accidents' are reported, but these refer to damage to vineyards rather than to workers. One is, in fact, surprised to find that there are few accounts of fires, explosions or poisoning. Later large machines were used to pump a mixture of carbon bisulphide and water around the vineyards, as is shown in Fig. 8. The traction engine, belching fire and smoke, is attended with the greatest insouciance by the intrepid engineer; hands in pockets, he stands, calm and confident, surrounded with blocks of fuel for his open fire. The observant, however, will note that he has been careful to place his machine and furnace up wind from the carbon bisulphide drums—his sentiments apparently updating the famous Australian motto by some years! It seems probable that M. Vialla was hurt by an explosion and a certain amount of poisoning, for he was not able to accompany the official mission to Mas de Fabre on 28th August 1873 because of 'a slight wound and great tiredness due to some experiments made on the mode of action of CS_2, rendering it impossible for me to walk and visit M. Faucon until the beginning of September'.[66] In 1945 carbon bisulphide was used on a considerable scale in the island of Jersey to help extinguish the Colorado potato beetle left in the island by the Germans. A machine fitted with several hollow tines down which the liquid was pumped was drawn through the ground by a tractor. Sparks originating from the striking of the tines against flints frequently ignited the chemical, and it was necessary to have men with fire extinguishers following the machine ready to squirt out any flash fire that occurred. One feels that similar flashes must have happened in the vineyards.

The news of carbon bisulphide poisoning in Paris caused a certain amount of alarm in the vineyards. However, at the Bordeaux Phylloxera Congress of October 1881,[24] a M. Géraud, of Bergerac, assured his audience that fears for the safety of vineyard workers exposed to the chemical were unfounded; making this product and using it in vineyards were quite different from using it in the vulcanizing process. One hopes he was right.

It soon became evident that this treatment was unlikely to extinguish the phylloxera, but that it did put a considerable obstacle in the way of its further progress, and large quantities began to be used.

The first method of treatment was for one man to make a hole with an iron bar, for a second man to pour in the correct amount of the liquid and for a third man to close the hole with his foot. Obviously this was laborious and expensive when it is considered that around thirty thousand holes had to be made per hectare. In addition the dosage was not very accurate. The man applying the insecticide poured in about the right amount each time; he could not be expected carefully to measure out, say, 10 cc. thirty thousand times per hectare! Soon simple hand machines were being made to facilitate this work; they were known as *pals*, a word difficult to translate. The first apparatus was a pointed iron bar which made a hole and was withdrawn, whereupon a hollow tube was inserted and the measured quantity of insecticide poured down (*pal*, David et Delbez). Next an automatic measurer was made (*pals*, Gayraud and Rousselier) allowing the measured amount to run down the tube; finally the injection principle was adopted. The first, and most popular, of these injectors was the *Pal Gastine*, but another, still made and in use to this day, was the *Pal Vermorel* (*see* Fig. 5). Two handles and a foot rest enabled the hollow tube to be pushed into the soil. The tube had a side exit and a foot valve. At the top was a reservoir for the carbon bisulphide and through it passed a piston pump operated by a central knob: the knob and attached piston being returned after use by a spring. The quantity delivered at each stroke was controlled by placing rings of differing thickness on the pump rod, thus adjusting its maximum travel to the required quantity. These machines were admirably simple and robust. Usually a team of two worked them, one man inserting the tube, delivering the dose and withdrawing the apparatus, and the other closing the hole and delivering fresh supplies of insecticide. A regular quincunx pattern at 60 to 80 cm. distance and 30 to 40 cm. depth was considered the ideal, but no injection should be made right against a vine stem. It should be noted that a quincunx distribution in equilateral triangles of 60 cm. sides gives a theoretical

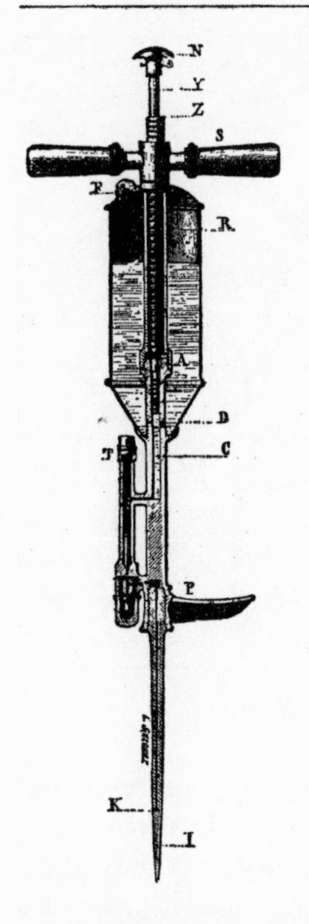

5. *Vermorel injector or* pal.

A *pump body*
C *chamber*
D *entry hole for liquid to the pump*
I *point*
K *delivery hole*
N *operating button*
P *footrest to assist in driving injector into soil*
R *reservoir containing carbon bisulphide*
S *handles*
T *valve mechanism*
Y *pump stem*
Z *ring regulating travel of stem; various sized rings can be fitted*

55,555 injections per hectare and even an 80-cm. triangle gives over 31,000; both are very large figures, and in practice from two to four injections per square metre (20,000 to 40,000 holes per hectare) were given (*see* Fig. 6).

Obviously this work is hard, long and expensive, and soon horsedrawn automatic injecters were being used. At least three kinds were made, those of M. Gastine, M. Vernette and M. Saturnin.

As the phylloxora crisis grew and wine production dropped, the Paris, Lyon, Marseilles railway company began to lose freight, and they devoted much thought and research to the problem, coming out eventually in favour of carbon bisulphide treatment. They based their work on some experiments of M. Allies. Foëx [75] points out that in the first tests the quantity of chemical to be used was based on insecticidal considerations and the plant was neglected. Vine roots were particularly susceptible to liquid CS_2, the subsequent evaporation leading to cooling and

6. *Using* pals *in a vineyard to inject carbon bisulphide.*

death. The P.L.M. started by recommending 30 g. per vine and then suggested the use of 15 g. per square metre in three holes, repeated at an interval of five or six days (*see* Note 13) thus using 300 k. per hectare. Later they realized that much smaller quantities could be used, until finally M. Crozier issued a neat little table showing four different soils each with five different depths and columns for three different states of the plants. Here the dosages ran from 110 to 280 kg. per hectare. In practice 150 to 250 kg. were used.

Carbon bisulphide was also used for the *destruction* of infected vineyards at the time when it was hoped that the pest could be wiped out entirely, and later to extinguish new foci. It became popular in Switzerland, Algeria, Germany and Russia when attacks were first noted. Very large quantities of the chemical were used, about 300 g. per vine, a ton and a half per hectare! It was used only for small patches and, the Swiss said, usually killed everything, the vine, weeds, worms, snails, spiders and all other insects. Even so, a few vines would occasionally shoot again, being found usually on the edge of the treated area and 'always free of phylloxera'. But, of course, though cured, the vines were always exposed to reinfection from the very moment of their cure.

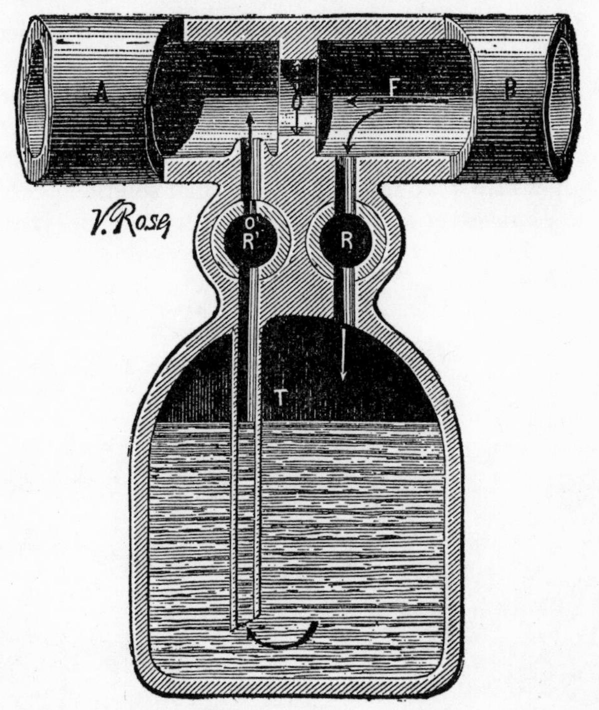

7. *The Faveur valve, used to mix small amounts of carbon bisulphide with a water flow in a pipe.*

The success of carbon bisulphide treatment depended very largely on getting an even distribution through the soil, and in 1875 it occurred to Professor Cauvy, one of Planchon's colleagues, that, since the chemical was slightly soluble in water, soaking the land with a solution of it would ensure this. The idea was followed up in 1882 by M. Rommier, a delegate of the Academy of Science's anti-phylloxera body, who suggested that whilst 2 g. per litre would dissolve in water it would be safer to use no more than 0·4 to 0·5 g. per litre. The main difficulty was to make large quantities of such a very dilute solution, and another was to find the large quantity of water needed. Making solutions in tanks provided with paddle wheels was not satisfactory, and the system was not much used until, after a delay of four years, the Faveur brothers' injector was invented. This simple device acted on the vortex tube principle (*see* Fig. 7). The chemical was automatically injected into the waterflow at a rate depending on the flow itself. Similar apparatus is used to this day to provide predetermined strengths of liquid fertilizers to crops. The water supply problem was another matter. For instance, theoretically, if a hectare of vineyard was to get 150 k. of CS_2, from a solution of a strength of 0·5 g. per litre it would need 300,000 l. of solution per hectare, equal to 3 cm. (over an inch) of rain. Very few vineyards were

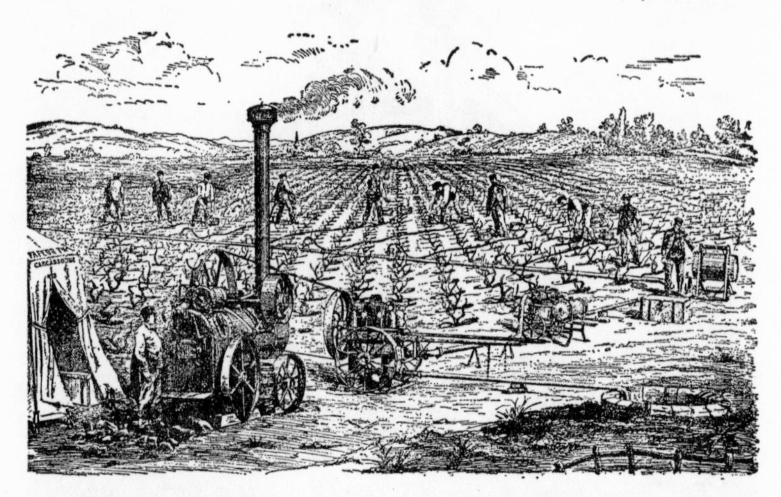

8. *Treating with water and carbon bisulphide. Note the* insouciance *of the engineer: note also that he is to windward of the highly inflammable insecticide!*

able to find such large amounts of water. In practice about 160,000 l. per hectare were used and rather stronger solutions, 0·6 to 0·8 g. per litre in winter and 0·4 to 0·6 g. per litre in summer. Both hand and steam pumps were used (*see* Fig. 8) and although the system was successful, Valery Mayet by 1890 [124] reluctantly declared that the French vines on their own roots could not be saved in this way but only have their lives prolonged a little beyond the fatal term set by the phylloxera. The peak usage of carbon bisulphide was about 1888, when 68,000 ha. were treated, needing some 10,200 tons of the fumigant. The price at that time was about 0·75 fr. per kilo or, say, a total of 7·65 million fr. (£330,000), a welcome addition to the chemical industry's sales but a severe cost for the *vignerons*. In fact the cost was probably considerably more, say 10 million fr., for many vineyards had two treatments per year.

It was found necessary to follow up CS_2 treatment with applications of fertilizer and better general care in order to give the root system a better chance of re-establishing itself. This meant that the advisers on phylloxera control appeared to be interfering in the matter of the general running of the vineyards and there was much opposition to them. In fact a battle was in progress between the phylloxera and the vine. The fumigant gave the plant a temporary advantage which had carefully to be followed up. Weeds had to be kept down, fertilizer applied, shoots tied in and the best of every routine attention given to the vine if it was to survive. Naturally there were many cases where the battle was lost and the vineyard started to perish, usually through neglect of aftercare if not of faults in the application of the insecticide. In opening the preliminary papers of the Bordeaux International Phylloxera Congress of 1881 M. Plumeau (and five colleagues) [24 44] say that they have to insist that much greater after-care be given to the vineyard than is usually the case, and on the whole they find *vignerons*, when told what they should do, '. . . as touchy as poets accused of writing bad verse'.

Carbon bisulphide could cause damage if used in too dry or too wet soils, or in too great quantities. At times, particularly in Burgundy, it was blamed for everything that went wrong, and sometimes the gangs ordered by a local committee compulsorily to treat a vineyard had to withdraw '*devant l'excitation des populations*'.[107b]

Carbon bisulphide was really the most practical of the insecticide solutions to the problem; it was welcomed by the Commission, composed largely of chemists who tended to believe in a chemical solution, although a biological method ultimately solved the matter. But it was difficult to persuade many people of its advantages, as the *Feuille Vinicole de la Gironde* of 1st October 1880 points out.

At the Bordeaux International Phylloxera Congress of 1881 much was made of this chemical, and the Viticultural Association of Libourne were its leading champions. A gold medal was awarded to the Société Marseillaise du Sulfure de Carbone for its remarkable stand showing the raw materials and finished products. A number of enamel, silver and bronze medals were given to other firms for chemicals and devices, such as M. Jousseaume's 'portable sulphuring furnace'.

Libourne seems to have been the centre of enthusiasm for carbon bisulphide, and M. E. Falières, secretary of the Libourne Association Viticole, its spokesman.[64] He quotes a number of 'well-known and satisfied users', such as Baron Thénard, MM. Monestier, Crulas, Rohart, Allies, and calls on his readers not to propagate puerile fears about the use of carbon bisulphide, but to make it better known and particularly to spread knowledge of the correct way of using it and the precautions to be taken. It cost, said M. Falières, 150 to 200 fr. the hectare and soon would be used by everyone, but it had to be followed by generous manuring and spraying against the winter egg, a touching belief in which was still quite strong in 1877. Carbon bisulphide could cause fires; so could many other products currently used, paraffin oil, for instance. M. Prosper de Lafitte, a 'winter-egg' enthusiast, aided by M. Rohart, made a certain amount of propaganda against the direct use of this particular chemical. In 1878 he gave a striking demonstration of its powers, properties and dangers. 'It is deadly,' he said 'for both cold-blooded and warm-blooded animals. It quickly catches fire' (here M. Rohart produced a small explosion). Calm being restored, M. de Lafitte brought out a live bird in a glass box, put a few drops of CS_2 on a bit of sponge on the inside of the lid and the bird was soon dead. The test was repeated with insects. 'You can see it is dangerous stuff,' he said. 'On the other hand combined with gelatine or combined in sulphocarbonates it is safe and effective.' M. Rohart was the supplier of the *Cubes Rohart*, a combination of CS_2 and gelatine. The idea was for the cube slowly to release the chemical into the soil and secure lasting protection. They did not prove to be of much use. Absorption on wooden blocks was also used.

Carbon bisulphide certainly had its successes, but the cost of 150 to 200 fr. per hectare seemed optimistic, for the necessary manures were not included in this price. M. Plumeau and his colleagues [24] thought that the total cost would be at least 450 fr. per hectare the first year and some 300 fr. in subsequent years, it being impossible to say if and when this expense would end. It was a cost fine wines could support, but not the *ordinaires*, especially if the price fell to 15 or 20 fr. the hectolitre.

Some of the failures of carbon bisulphide were due to roguery on the part of contractors or workmen who could bypass the pump on some *pals* and appear to be injecting the fumigant when they were not. Farmers were advised to supply their own carbon bisulphide, when the contractor would have no temptation to economize on its use, or to use an injector, such as the *Pal Gastine*, where this could not be done.

CHAPTER TEN

Other Remedies

SULPHOCARBONATES

Carbon bisulphide was evanescent. It killed the phylloxera and then was gone, either into the air or into drainage water, and the vine was back at stage one again, ready to be infected by any passing pest. Moreover it was difficult to ensure the even distribution of the gas in the soil.

M. S.-B. Dumas, the famous permanent secretary of the Academy of Science, conceived the idea that the alkaline sulphocarbonates would be a good substitute for carbon bisulphide, and he brought the matter up in June 1874.[61] Sodium and potassium sulphocarbonates exposed to CO_2 gas and water, in air or soil, slowly decompose and liberate carbon bisulphide and H_2S gas, both poisonous to the pest. It was thus thought that the insecticidal effect would be longer lasting than that obtained from the direct application of carbon bisulphide. As the salt absorbs water rapidly, it was usually sold as a solution of 40° Baumé gravity, containing 55 per cent of pure potassium salt. 100 g. of this solution on decomposition would thus give 20 per cent of CS_2 and 9 per cent of H_2S, say 6 l. of each gas. Some tests were started at the experimental station set up by the growers themselves in the Cognac, by two delegates of the Academy of Science—MM. Cornu and Mouillefert. Wisely they started with laboratory tests and found that the insects could be killed in from fifteen minutes to a day, according to strength used and time of exposure.

These encouraging results led to field experiments, also very successful. In order to get the chemical into the soil at all it had to be watered in, and to get it well distributed much water had to be used, which again was the fatal drawback to the wider use of this method. Though it had the advantage of a greater safety margin than carbon bisulphide it was more expensive: 400 to 500 k. of sulphocarbonate per hectare were used in 100,000 to 150,000 l. of water, equal to a centimetre of rain at the lower figure. MM. Mouillefert and F. Hembert devised a system of portable steam pumps, hoses, tubes and so forth in efforts to cheapen and simplify the application, but even so it proved costly and, as before, the vineyard needed the assistance of fertilizer to help it recover. Both sodium and potassium sulphocarbonates killed the pest; the sodium salt was cheaper, but many vineyards seemed to be short of potash, so that it was well worth using the more expensive chemical as it gave a boost to the vines after the insect had been killed and, by being in solution, the potash reached the exact spot needed. M. Plumeau and his team estimated the cost of treatment as from 400 to 500 fr. the hectare, and 600 fr. even, with another 200 fr. for fertilizer. As the chemical sold for about 1 fr. to 1·20 fr. per kilo they thought the cost could be taken at 700 fr. the first year and at 500 fr. for all subsequent years. M. Mouillefert, the sulphocarbonate enthusiast, disputed these figures.[132] A company had been formed (la Société nationale contre le Phylloxera, apparently directed by M. Mouillefert), which undertook contract work. In 1876 they had carried out 173 treatments over 1,200 ha. in the south-west, and payments made by owners worked out at 345 fr. per hectare on 517 ha. and 327 fr. in the Midi on about 630 ha. or an average of about 335 instead of the 500 mentioned by the Phylloxera Commission's *rapporteur*. Sometimes the cost went below 200 fr. and never reached 600 fr. This suggests they had started making sulphocarbonates on a big scale and had brought down the price. These figures did not include the cost of manure which, M. Mouillefert maintained, should not be entered because the vines ought to be fertilized and fertilizer was thus a normal cost; nor did they include the cost of striking channels, which were merely another working of the soil, needed in regular culture of the vine in any case. Low figures were obtained, they said, because the company used an experienced labour force and worked on a big scale.

Vignerons ought to get together to form anti-phylloxera groups and keep down their costs. Obviously this, whilst true, would also facilitate M. Mouillefert's business. Later on in the Bordeaux Congress [24] (1881) the Comte de la Vergne thought M. Mouillefert was dressing the shop window with certain show vineyards in order to establish his system and his trade. Nevertheless he (the Count) had obtained good results with sulphocarbonates and carbon bisulphide. The Count said that people should be grateful to the pioneers, to M. Monestier, for instance, for taking up carbon bisulphide again after the legendary Thénard 'disaster'. Weight is added to the window-dressing theme by an account found in the Château Loudenne (Gironde) archives. On 24th April 1883 the Château paid the Société nationale contre le Phylloxera the sum of 15,078·80 fr. for treating 301,576 vines, said to be 34 ha., hence costing 443 fr. per hectare, considerably more than the figures quoted by M. Mouillefert.

In the 1880–1 season the company was 'very active' and treated 1,163 ha. (five and a half million vines) and used 443 tonnes of sulphocarbonate. This suggests they were using 380 k. per hectare and had a turnover of some 390,000 fr., not an enormous business as yet. They had many satisfied customers. M. Henry Marès said that in his area, the Launac plain, at one time healthy and prosperous vineyards stretched as far as the eye could see. Now there were but 15 to 20 ha. alive; treated for the last three years with sulphocarbonate they gave 100 hl. per hectare in spite of numerous missing plants. 'Sulphocarbonate,' he said, 'sews the grapes onto the vine.' [119]

The relatively high cost and the difficulty of finding water restricted its use to a comparatively small area, the maximum in any one year being 9,000 ha. compared with 62,000 for carbon bisulphide.

PLANTING IN SANDY SOILS

The root form of the phylloxera is unable to thrive in sandy soils. It could easily be imagined that sand would facilitate the movement of a living insect from root to root in soil, but the contrary appears to be the case. The phylloxera thrives in clayey soils and escapes from its habitat as the soil dries and cracks in summer; in sands it apparently cannot move after it has killed a root. No cracks last in such soils since they fill by the easy movement of the particles as soon as they open, and imprison any phylloxera attempting to emerge. This surprising feature was first drawn to the attention of the world by a small *vigneron*, M. Ch. Bayle; the phenomenon was studied by two scientists, MM. Marion [120] and Vannuccini.[177] They found it to be true and put it down to the easier penetration of water into sands, especially those near the Atlantic seaboard, where high tides would raise and lower the water table. By 1874 M. Espitalier [62] had issued some practical suggestions for growing vines in sands and new areas were planted.

It was found that to resist the pest the soil must consist of at least 60 per cent siliceous sand and that calcium sands (shell sands) were not as good in this respect as the former. Vines throve in most sands except those having too much sea salt. In spite of the presence of the pest in the area vines were growing well in the Gascony *landes* and the Gulf of Lyons, particularly at Aiguesmortes. They were later found to grow well on their own roots on the sandy Tunisian coast and the Algerian Sabel. Some alluvial river sands in the Rhône valley also proved capable of supporting own-root vines.

Sands are usually poor soils, and if fertilized with compost and farmyard manure can be much improved. This altered their nature it was found and, as the organic matter in the soil increased, rendered the vines liable to attack by the pest. In other words the soils were ceasing to be sandy. Chemical fertilizers were not so dangerous in this respect. So advantageous was this discovery that certain sandy areas, particularly those around Aiguesmortes, increased in value and were planted out. A drawback quickly arose; the soil tended to blow about. Vine roots were exposed and dried out; vines were buried and even dunes formed and moved. This was overcome by *enjoncage*, the planting of lines of reeds between the vines to fix the soil. M. Vernette, of Béziers, even made a special machine to facilitate this planting. It consisted of four iron discs on a framework which, drawn through the sand, opened up four narrow furrows into which bits of rooted rushes could be placed. The efforts which growers would make to preserve their cherished vineyards may be seen in this, for about a thousand rush plants per hectare were needed. Another disadvantage was exposure to storms and the blowing in of salty spray from the sea, which burnt the vine foliage.

According to Foëx,[75] on the better Aiguesmortes sands,

previously cultivated for madder crops, high yields of 250 hl. of wine per hectare were obtained, but another writer, M. Aguillon, says 80 to 100 hl. were average and manuring could raise it to 150 hl. Prices of land rocketed in this area. Dunes selling at one time for 100 fr. the hectare now made 3,000 fr.; workers and former proprietors flooded into the area (and to Vauvert and Générac) to plant new vineyards and somehow contrived to exist for seven years until the new plants were in full production. In fact the area accounted for a good part of the 8,000 ha. of vines still remaining of the former 95,000 ha. in the Gard. By 1883 there were 5,000 to 6,000 ha. of vines in the Aiguesmortes area alone.

Another such sandy area was Sète and Agde in l'Hérault which very much prospered, due to their immunity, except at Onglus, where the vineyards were tiny and the proprietors in the hands of money-lenders.

It was thought that vineyards could be saved, or at least have their existence prolonged until a remedy was found, by excavating the earth around each vine and replacing it with sand. Lichtenstein,[115] for instance, reported that he had restored productivity to 80 per cent of the average by this process. In the Camargue, M. Espitalier saved vines in this way, using 80–100 l. of sand per vine, plus 250 kg. of guano and a tonne of ashes per hectare. That was all very well, but sand weighs about 1·343 kg. per litre; a hectare of say 5,000 vines would thus need from 535 to 770 tonnes of sand per hectare, an impossible quantity, as was soon learned in practice. Even so, 100 l. is not much of an area for a vine's roots to exploit, it is a tenth of a cubic metre and even a cubic metre is not a large volume of soil for such a plant. However, where sandy soils were available and not too salt, own-root vines survived and do so in considerable numbers to this day.

A BATCH OF MISCELLANEOUS SUGGESTIONS

A number of other sensible suggestions were put forward. Many of them relied on the production of sulphur dioxide (SO_2) or sulphuretted hydrogen (H_2S), both of which gases were insecticidal. Many others appeared to be successful because they checked an incipient invasion and supplied sufficient fertilizers to enable the vine to recover at least to some extent, and, clutching at any straw, the experimenters were persuaded they had a remedy.

The processes below are arranged approximately in date order and are a selection from among the hundreds tried in one place or the other. In 1873 *The Times*[174e] reports that littering the vines to the depth of a foot with green tobacco plants provides sufficient nicotine to kill the pest. Obviously the remedy was quite impractical, for a hectare of vines would have needed at least two hectares of tobacco to provide the litter. In 1874 M. Elie de Beaumont found snow piled round the plants was a cure. In 1875 a correspondent in *The Times* recommends aesfoeteda as a cure, pointing out that it had cured a similar pest in India. In 1876 M. Rohart, knowing that CS_2 would kill the pest but leave the plant exposed to reinfection, proposed to bury a glass flask containing this substance, one having a special closure to allow the slow release of the fumigant. It was not successful. Several remedies appear in 1877. One scarcely worth mention, except that it was supported by Millardet (later the inventor of the cure for downy mildew—Bordeaux mixture), was *Poudre Garros*.[24] This arose from the phylloxera/effect school, for the powder healed the pricks in the vine roots made by the phylloxera aphids and prevented the entry of the damaging fungus. It thus, according to M. Millardet's happy phrasing, 'Americanized the French vines' and rendered them immune to attack. In the same year a 'Mozambique Oil' was patented. It was said to be fish oil in which anti-helminthic plants (intestinal worm remedies) had been steeped. Obviously it had but little effect on the pest, but it achieved great fame in the Bouches du Rhône whose horticultural society gave it a silver medal. A litre cost but 2 fr. and treated 100 to 200 vines. Cheapness must have been one of its attractions, for it came out at 50 to 100 fr. per hectare compared with 600 fr. for sulphocarbonates. One wonders what it had to do with Mozambique. Perhaps the name brought in a suggestion of African magic. Two professors praised it, MM. Heckel and Sicard.

Around this time further SO_2 processes aroused interest; M. Mouline made modest claims for his 'sulphured charcoal'.[24] He found that charcoal would absorb up to sixty-five times its volume of SO_2, give this off when mixed with soil and kill the pest. The difficulty, naturally, was the incorporation of the product in the soil. In 1877 MM. Chalbrul and Lunel took up an idea, previously put forward by MM. Haltz and Duluc, for blowing burning sulphur along drainage pipes run through the vineyard. They claimed that the pipes would cost no more than 0·12 fr. the metre run, that they would last for from fifteen to twenty years and sulphur would only cost 12 fr. the hectare. It was a modest sum for pipes. Even so it comes to a capital outlay of 680 fr. per hectare and vine roots would have blocked the pipes long before even fifteen years were passed. To overcome the piping difficulty M. Fusélier of Angoulême invented a fearsome machine—the *charrure-fourgon-chaudière*. A furnace and boiler fed fumes from burning sulphur and 'dried herbs' down a tube into the plough furrow, between the coulter and the share; the fumes were thus to be trapped as the slice was turned, killing the pest. Reports from the Ardennes praised it and the extra cost was but 2 fr. per hectare, as the land had to be ploughed in any case. It seems extremely modest and cannot have allowed for any cost of the machine itself. It is doubtful if enough SO_2 could have got into the soil in this way or would have been sufficiently evenly distributed. Sufficient SO_2 to kill the insect was likely to have killed the vine too.

M. Boué of Meilan (Lot et Garonne) used potassium sulphide, wood ashes and lime to release H_2S in the soil, and M. Eymaël, of Liège, employed an alkaline sulphide and aluminium sulphate for the same purpose. Both these processes had the disadvantages of needing a large quantity of water, and M. Eymaël advised treatment when the barometer was low or when advised from America of a coming storm, as abundant rain would distribute the chemical in the soil. Presumably the Atlantic cable was used for weather forecasting at that time.

M. Vigié of Marseilles designed a machine to inject hot SO_2 at the base of each vine. It was a somewhat naïve apparatus, consisting of a bellows and a burner into which the operator dropped a teaspoonful of flowers of sulphur from time to time. It had a long spout which was pushed into a hole made previously by an iron bar. Over a period of four years, repeated two or three times a year, it did no harm to the vines and whilst it did not completely destroy the pest it kept it within bounds. According to M. Vigié it also controlled the oidium.

M. Maurin, of Toulouse, was a believer in the winter-egg theory and he proposed to break the cycle by destroying this with the *bergonette artificiel*, an instrument provided with a coarse brush, a soft brush and point used to scrape the bark, which was later

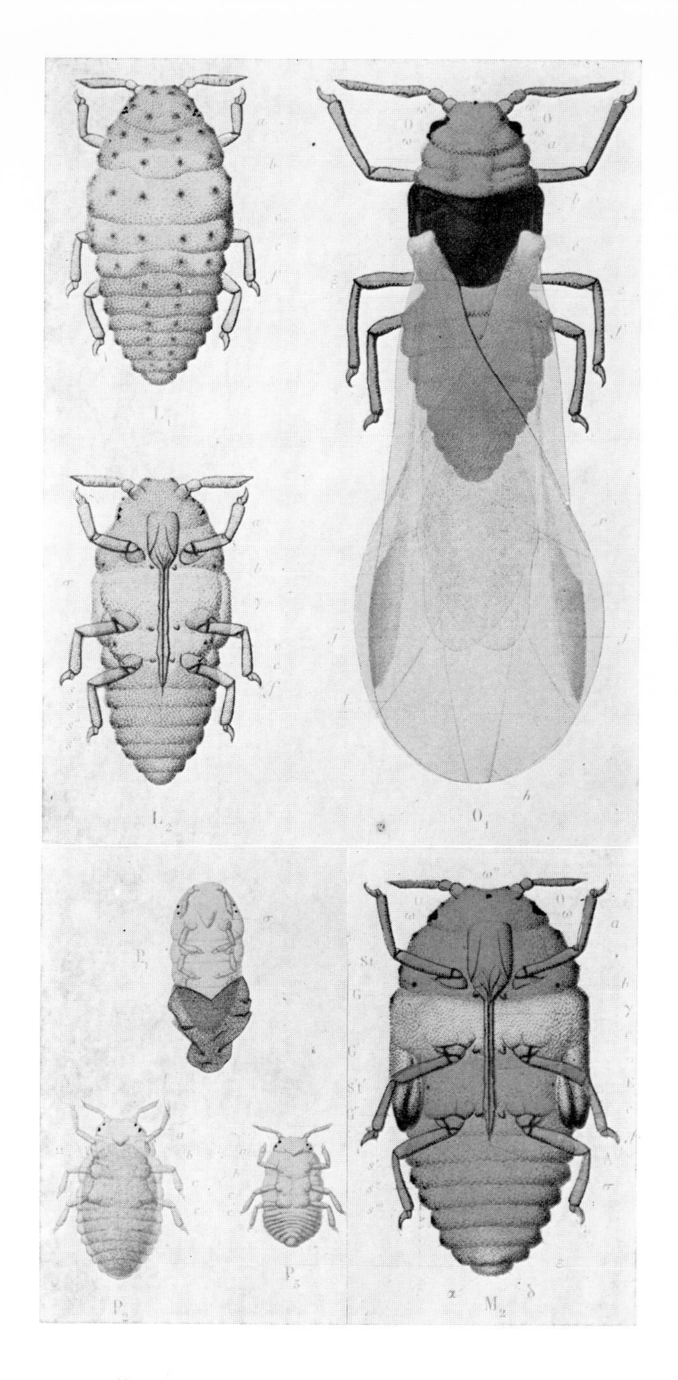

1. Phylloxera vastatrix

L_1, L_2. *Insect developing into nymph, dorsal and ventral view;* M_2, *nymph, ventral view;* O_1, *winged female, dorsal view;* P_1, *insect hatching from egg;* P_2, *female containing egg;* P_3, *male.*

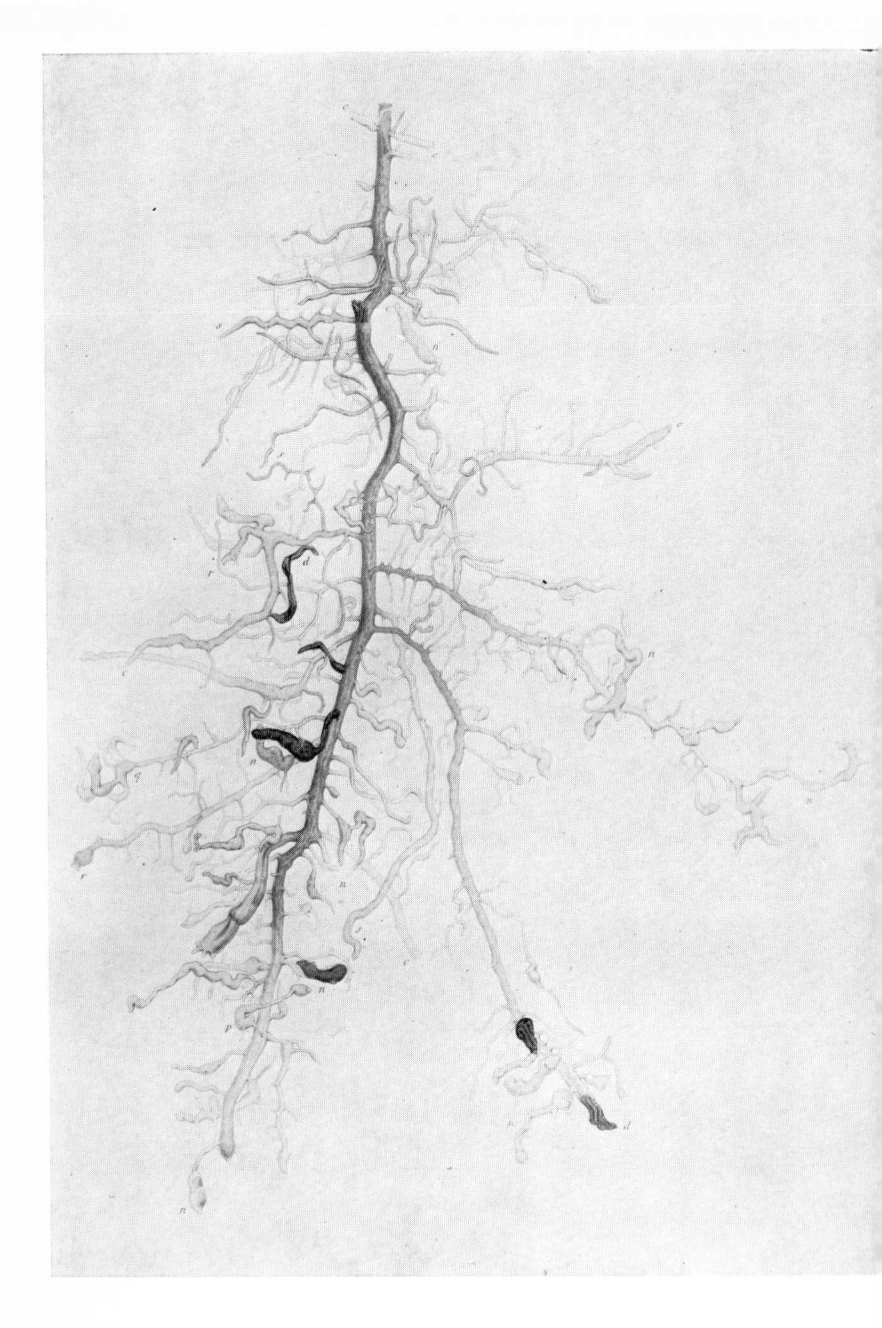

2. *Damaged* vinifera *vine root, showing swellings and dying rootlets.*

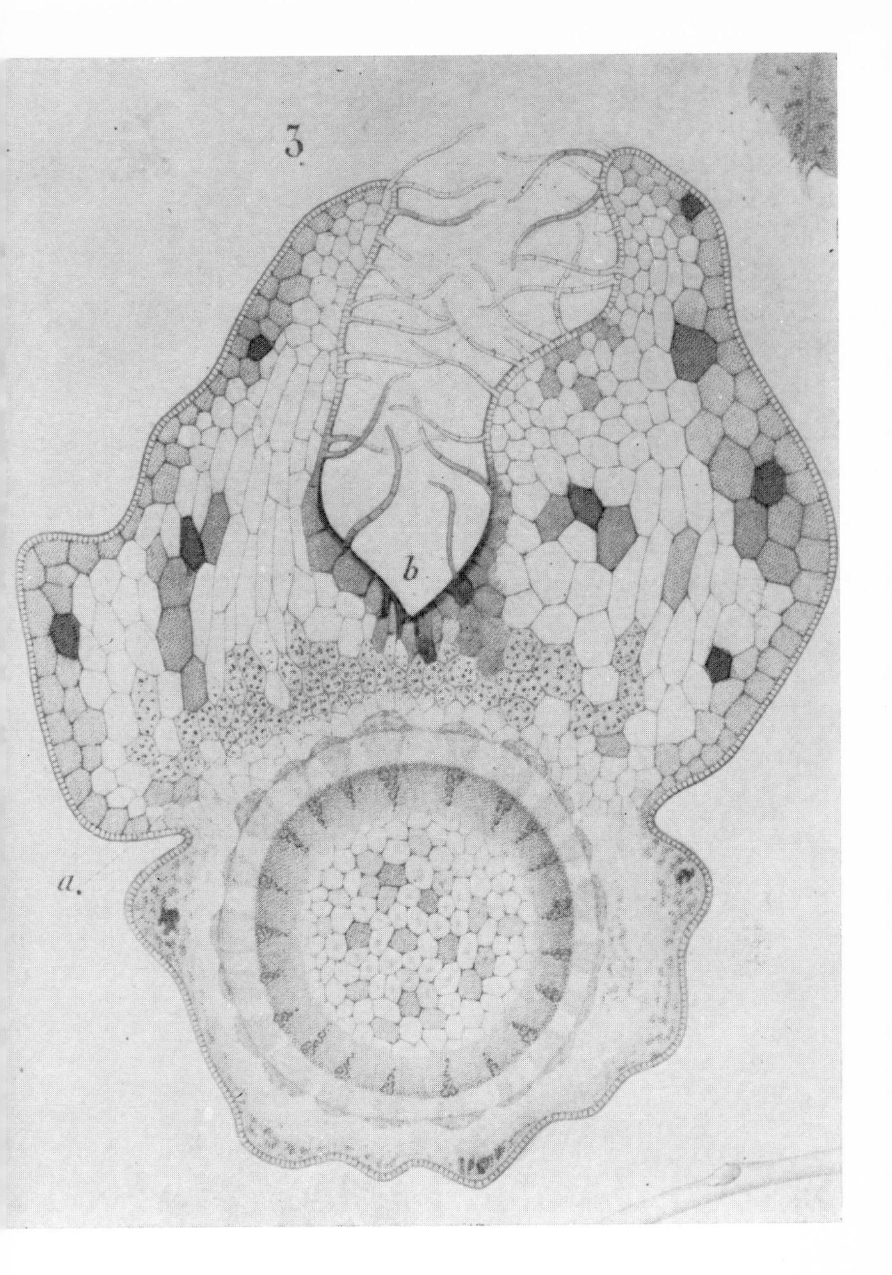

3. *Cross section of* Phylloxera *gall on a vine tendril. Note the hairs protecting the entrance to the gall. Approx.* × 50.

4. *Vine leaf with* Phylloxera *galls.*

5. *Inspecting an infected vineyard 1878. The top-hatted gentlemen are government inspectors (possibly MM. Planchon and his brother-in-law Lichtenstein), and the man in the 'wide awake' is probably C. V. Riley from the U.S.A.*

6. *An attacked vineyard near Sète, 1874.*

7. *Flooding of an 'own-roots' vineyard near Sète, 1970.*

8. *C. V. Riley.*

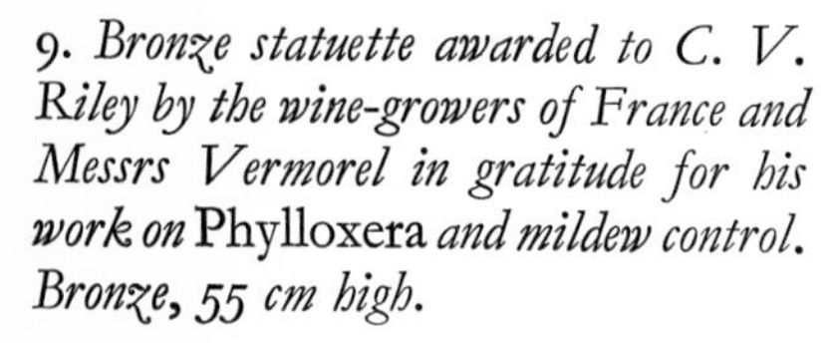

9. *Bronze statuette awarded to C. V. Riley by the wine-growers of France and Messrs Vermorel in gratitude for his work on* Phylloxera *and mildew control. Bronze, 55 cm high.*

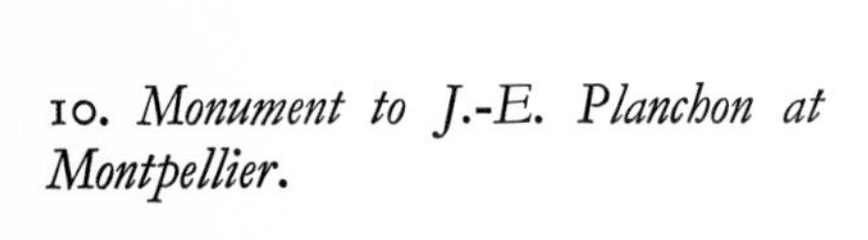

10. *Monument to J.-E. Planchon at Montpellier.*

11. *Memorial at the School of Agriculture, Montpellier, France. The ailing, old, distressed French vine is being rescued by the nubile young* Américaine. *Note the date, 1911: not a product of the modern 'permissive society'.*

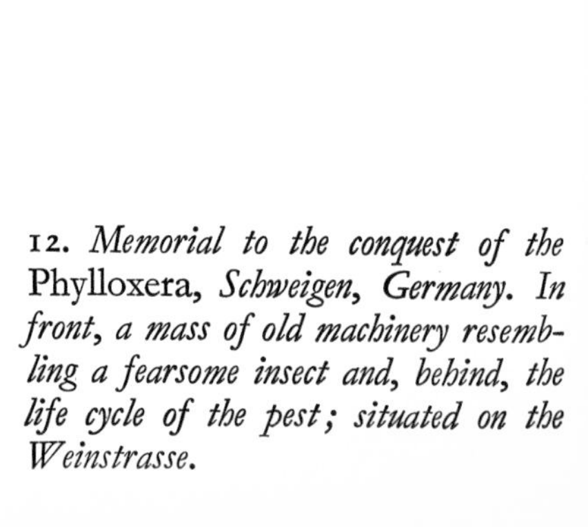

12. *Memorial to the conquest of the* Phylloxera, *Schweigen, Germany. In front, a mass of old machinery resembling a fearsome insect and, behind, the life cycle of the pest; situated on the Weinstrasse.*

10

11

12

13. *Lantern lecture on the pest, Paris, October 1874, a fashionable audience in attendance. It was the first time that electric light was used for this purpose.*

14. Phylloxera *was joked about in th national press.*

— Éloignez-vous, monsieur! le jus de la treille dans ce-nez là! le phylloxera va vous sauter à la figure.

washed with soap and carbolic fluid. Even had the winter egg existed, destroying it would have made no difference to the root-living forms really doing the damage. The treatment was useful against two other pests though, *Cochylis* and *Eudemis*.

Dr Mandon, of Limoges, foreshadowed our modern systemic insecticides. He maintained that a 1–500 solution of commercial phenol in water killed the phylloxera and that the vine would take up such a solution, transmit the chemical to all parts and kill the pest wherever it was. The doctor found himself up against one of the problems of modern systemics, that of residues and taint. Treatment had to be stopped a month before harvest or the grapes would carry a carbolic flavour. The method was praised at one time, but modern tests with phenol absorbed by plants (*see* Note 10) show no aphicidal action and we think the doctor's good results must have been exceptional, for the method was never much used as, being so simple, it would have been had it worked.

In 1877 insecticidal fertilizers were advertised. *Le Vigneron Champenois*[184] carries one for the Procédé Mallet-Chevalier containing a 'potassic naphthate' and minerals and costing 240 fr. the ton with 2 per cent discount. Agents were being sought.

In 1881 Major-General C. L. Showers wrote a long letter to *The Times*[174h] pointing out that the pest was due to the overtaxing of the vines and the remedy was to take smaller crops. Obviously he had very little idea of the true situation. The same year the paper[174i] carried a report of Signor Bourbon's 'fire-bellows' in Italy, seemingly a flame-thrower to be used against the supposed winter-egg.

On 5th September 1894 an enterprising woman took out a patent for an anti-phylloxera preparation and was awarded a medal by the local Comice Agricole. She was Mlle Tourné, postmistress at St Macaire, Gironde. She says she experimented for five years with her insecticidal fertilizer, made as follows (for each plant): water 1 l., soot 200 cc., oil 3 g. and sulphur in powder 330 cc. or 9 g; 200 cc. of sawdust might also be added. The preparation was applied to the roots of each plant 10–15 cm. below the surface. It was unlikely to have had any effect on the pest, and her early success must have been more due to luck than anything else. About this time the placing of a *volcan de Lémeny* at the base of each vine was seen as a remedy. The *volcan* was a mixture of iron filings, sulphur and water (invented by Nicholas Lémeny, a French apothecary, born in Paris in 1645) which when gently heated gave iron sulphide with the liberation of heat and 'vapours'. M. Lémeny maintained that this was how volcanoes formed and erupted and used to demonstrate his *volcan* to the wonderment and delight of spectators. Some of the vapours would have been SO_2 and would have had an insecticidal action, but the remedy was hardly practicable and, as with all insecticides, left the vine, even if cured, exposed to new attack.

We close this chapter with an account of some of the more esoteric remedies proposed. Two kg. of ice around each plant was said to do the trick. Ten tons of ice per hectare, even if it worked, would need some finding in spring or summer. M. Apolis, of Montpellier, gave his views on both cause and cure which are remarkable more for their presentation than their content. M. Apolis thought that the vegetation had suffered from cooling, caused by a Miasma, leading to the plants becoming feverish and thus producing the phylloxera. In addition the soil might be sick and lead insects in it to abandon feeding on soil and turn to feeding on plants. M. Apolis's cure was Transpiration. He does not say how this is to be done, but presumably one had to write to him for further information. The presentation of this remarkable idea is on a large, handsome engraved sheet in a beautiful copper plate hand (one was sent to Dr Hooker at Kew).[106d]

Another explanation and cure was that of Mr George Davis, Professor of Physiology, of London, who said the cause was soil exhaustion and the cure his special fertilizer, 150 fr. the ton f.o.r. or f.o.b., London, special prices for larger quantities.[106c]

Vegetable insecticides were often thought of, a typical one being

Hop leaves	500 g.	cost 1·00 centimes
Mallow flower	250 g.	1·75 ,,
Orange leaves	25 g.	0·25 ,,
		3 centimes

Boiled in 60 l. of water and added to 600 l. of fresh water it was said to kill the pest when watered on the vines. Perhaps the water helped the vines in a dry summer (*see* Note 14).

Cornu and Mouillefert at Cognac[47] divided the substances

tested into seven groups and found nineteen among them with some activity against the pest. In the list below these active materials are marked *:

- I Manures: guano, super, urine, purin, marc, compost, farmyard manure.
- II Neutral substances: water, plaster, soot, coal, sand.
- III Alkalis: tar, ammonia, potash, soda, soaps.
- IV Salts: of iron, copper, zinc, potash, ammonia, sea salt, mercury bichloride, arsenious acid, alum, yellow prussiate *, potassium cyanide *, potassium sulphocyanide, calcium phosphate.
- V Vegetable products: chamve, *Datura* * (thorn apple), absinth, quassia, valerian, walnut shells and husk, oak bark, aloes, tobacco, spindle, aspic oil *, vegetable oils *, oil cake, olive oil press cake.
- VI Strong smelling substances: coal tar *, wood tars, paraffin *, benzine *, benzol *, carbolic acid, naphthalene, creosote, heavy oil *, aniline, schist oil *, asphalt, picric acid *, 'Vicat anti-p. insecticide', cade oil, turpentine *.
- VII Sulphur: hydrogen sulphide, sulphocarbonates of ammonia*, potassium *, sodium * and barium sulphides *, potassium and calcium sulphates, barium and calcium polysulphide *, iron sulphide, sulphur, sulphurous acid, sulphuric acid, aluminium and potassium bisulphites, ammonium sulphohydrates *, carbon bisulphide *, mercaptan.

The Commission Supérieure du Phylloxera was reorganized by the law of 15th July 1878 (it had originally been set up to adjudicate on the award of the 300,000 fr. prize) and in its 1879 report rather wearily comments: 'The names of the inventors change each year but the remedies they suggest are always of the same kind. As usual a few facetious suggestions slip in among a number of eccentric proposals put forward in good faith.' The Commission then goes on to enumerate some of these, but unfortunately does not distinguish those it considers facetious from those it holds to be eccentric. Perhaps the following mentions were both: grafting vines onto mulberries, blackberries and gooseberries; using oil, sulphur, sea water, tobacco, tar, phenol, coffee grounds, incense, intercrops as attractants or repellants, birds, ants, manures. All the known methods were reproposed as well as suggestions for biological control by a disease or parasite.

The definitive cure was grafting on to resistant American rootstocks, which is dealt with in the next chapter.

CHAPTER ELEVEN

Rescue from America

Certain American species of vines had been introduced to France in the eighteenth century, and many imports were made in the 1840s in the hope of finding a variety resistant to the oidium mildew. They did not bring the phylloxera with them because of the slow journey. It was soon found that the American grapes made very poor wine.

When the phylloxera struck and inexorably spread it was quickly noticed that certain vines escaped destruction and that these were always American. As the pest got worse and nothing seemed to stop it the idea naturally arose of growing only American species, and from that the notion of grafting the good-wine European grape onto the resistant American rootstock.

The idea of grafting the European vine onto American roots is considerably older than might be thought. Cortés finished his conquest of Mexico in 1521 and to consolidate the Spanish hold on 'New Spain' a large number of laws were imposed, regulating the development of the area and the treatment of the inhabitants. Particularly, this legislation imposed conditions as to how proprietors were to look after the *repartamiento* Indians (slaves in fact) and the land. A law of 1524 says that in order rapidly to obtain wine, owners must plant out as expeditiously as possible a thousand vines for each hundred Indians they hold until a total of a thousand established vines for each hundred Indians was reached. If this were not done the owner would be fined half a gold mark for failure in the first year, a mark for failure in the second and would lose the Indians if he failed again in the third year. Moreover, in order to get the vines to fruition earlier the Spanish cuttings were to be grafted onto established native plants (*see* Note 15). It is a little ironic that there was no phylloxera in Mexico, nor is there today, so that from the pest control point of view grafting was not needed, but there is no doubt that it must have hastened the establishment of vineyards in that country. Mexico thus provided no evidence for the importance of grafting. But an early example in the eastern U.S.A. also went unnoticed. Mr Buckley, the American botanist who studied the flora of Texas, has a record in the Philadelphia herbarium as follows: 'No. 79. Black sweet grape, introduced vine; flourishes finely when engrafted on a strong mustang vine, in favourably situated,* produce 10 bushels of grapes the fourth year: Will not thrive with its own roots without a great deal of careful nursing. Blooms in April, ripes [*sic*] in August.'

It seems to be M. Gaston Bazille who first suggested grafting, at the Beaune Congress of 1869, but later M. Laliman claimed priority, maintaining that he was the first to draw attention to the resistance to the pest of American species. Laliman later claimed the prize (*see* Appendix). Foëx examines this resistance with a great grasp of the theory of natural selection, a somewhat new and not very popular idea at that time. (Darwin was refused admission to the French Academy of Science in 1872 by 32 votes against to 15 for.) The phylloxera, he maintained, could not be a European insect for, if it had been one, it would have wiped out the vine as a species and then would have become extinct itself for lack of food. In America, over the centuries, the insect and the host species had adapted themselves to each other and American vines could live with the pest on them, even on their roots, a mutual tolerance to the advantage of plant and insect having been established, similar to *concordats* between Pope and Monarch and cops and robbers.

Foëx examined the matter from two points of view, one the

* The above is from Foëx's English and at this point has the words in brackets '(sic) *pour* unfavourably'. The English appears to have suffered in transcription. The author is seeking the actual wording of the Philadelphia herbarium entry, which has not been located at the time of going to press.

practical experience of a number of American vines actually growing in France, and, two, the scientific aspect, the former, seen from the present day, being rather better than the latter. He had come to the conclusion, at least by 1888 and probably much earlier, that it was no good trying to keep the French vineyards going by constant treatment, 'inevitably reduced to the condition of a sick man kept alive by constantly stuffing him with medicine' as he put it. American vines were the radical cure with two main possibilities, reconstituting the vineyards with 'direct producers', that is, American species or hybrids, or grafting the European vine onto American roots. In the first case an unusual and not as pleasant wine would be produced; possibly special methods of vinification would get rid of most of its unpleasant taste; in the second there were a considerable number of unknowns in the equation. How resistant were the Americans, and did resistance vary from species to species? Grafts had been made, but could they be made in millions and how long would they last? Finally, would the American stock pass the American flavour to the European scion and all the trouble of grafting be wasted? We now know the answers and put them in here, but they were very worrying questions at the time. Some American species were destroyed by the phylloxera (*labrusca*, for instance) and others were immune to damage from the root forms (*rotundifolia*, for example). A scale of resistance was eventually established. Grafts can be made on a big scale and grafted plants last twenty to thirty years, not as long as own-root plants, but long enough. The last question is still debated, but most people would agree that the stock has very little influence, if any, on the flavour of a wine (*see* Chapter 15).

That indefatigable researcher and advocate of American vines, the Duchesse de Fitz-James, even found some quite ancient grafted vines in France. M. Cazalis-Albut successfully grafted quite a large area in 1823 in order to change the variety quickly. In 1830 he succeeded in grafting an eighty-year-old vine and reported the matter to the Oran Congress in 1886. Presumably he was of considerable age at this meeting. Many large-scale trials and observations were made to test this idea. In addition to Laliman's plantation near Bordeaux, of some eighteen years' standing, there was another, of M. Borty, at Roquemaure, Gard (he probably introduced the pest as well; Roquemaure was an early infestation site) of about the same age. Both were healthy, though reported as being sometimes attacked by the insect. Probably this was on the leaves.

M. H. Aguillon, of Chibron, near Sigues (Var), lost nearly all his vines to the phylloxera and decided to replant several hectares with vines of many kinds. He set out 150,000 cuttings of 800 different varieties at the beginning of 1872. This was an enormous and most enterprising undertaking. The cuttings took well, but by the second year the local varieties (the *viniferas*) started to fail and were soon all dead. Only York-Madeira (a hybrid), Jacquez, Cunningham and Herbemont (*æstivalis*) and Taylor (*riparia*) persisted and were still there twelve years later, in spite of being in poor, stony soil.

M. Reich, of Armeillères, near Arles (Bouches-du-Rhône) was even more determined to get to the truth than M. Aguillon and put his Americans to a severe test. In 1875 he dug out a big area and covered the bottom with roots attacked by the pest, returned the soil and planted American and French vines over them. None of the French survived whereas most of the American ones throve, the exceptions probably being the *labruscas*. As late as 1888 people were still repeating tests of this nature and one can sympathize with them; so many 'cures' had been announced and had failed, the task of replanting a vineyard with entirely new and unknown varieties was an enormous expense, and, finally, it was hard to see why an insect which killed stone dead a French vine should really leave a very similar, but American, plant quite unharmed. The exact reason for this difference is hard to see to this day; a few people are working on it.

From the kind of tests described above, and from general observation and advice from Mr Riley, it was soon found that certain American vines did resist the pest, particularly in the Hérault, where the attack was worst, so that growers soon faced up to the problem's only practical solution. The American-vine area rose from 2,500 ha. in 1880 to 45,000 ha. in 1885.

The rescue from America was greatly helped by Planchon's visit there in 1873. An extremely conscientious and reliable man, he brought back first-hand information of how American species grew, though, in point of fact, less was known in America about grafting and root resistance than in France. The journey had great psychological value on his return: here was a man who knew the subject, had been to America, had seen the vines there and he thus talked with authority for the Americanist school. In the U.S.A. Planchon was much helped by C. V. Riley, who gave up a planned botanizing expedition in the Rocky Mountains to meet Planchon off the boat in New York. D. W. Morrow, Jr [131] has assembled an interesting account of Planchon's views on the United States at that time. In a five-weeks' rush the two scientists visited New Jersey, Pennsylvania, Baltimore, Washington D.C., North Carolina, Ohio, Missouri, western New York and eastern Massachusetts. Mr Morrow has examined Planchon's personal notes on his tour (now in the School of Agriculture library, Montpellier) and says that he (Planchon) obviously intended to write them up for some journal as a travel article. Possibly this was the *Revue des Deux Mondes* to which he contributed, but the article never appeared.

Planchon much enjoyed his visit to the French colonists at Ridgeway, North Carolina, where he was made very welcome and billed as one of the attractions of the State Fair, though it was due to take place well after his departure. He admired the boldness and achievement of the American railways and also the city of Cincinnati, where he met and discussed vines with Robert Buchanan, the author.[31] Planchon comments on the absence of cafés in America, which may be good for morality but is bad for sociability, and he notes that ladies go to ice-cream saloons, and gentlemen too, though they were not meeting-places for conversation.

Unfortunately the vines Planchon and Riley recommended, whilst resistant to the pest, did not suit the majority of French vine soils, which were mostly chalky, and it needed another visit, that of Pierre Viala, to straighten out this difficulty and find suitable rootstocks.

The scientists were naturally curious as to the cause of the difference in reaction to the phylloxera between the species, and several possible reasons were advanced and rejected. At first it was thought that the American plants were so much more vigorous in their root growth, particularly in making new fibrous roots. This theory could not stand up because new fibrous roots were just what the phylloxera liked, and a closer examination showed York-Madeira growing well with a poor root system and Aramon (a *vinifera*) dying with an extensive one. In 1876 M. Boutin put forward the view that the American roots held resinous substances that blocked the feeding holes made by the phylloxera insects, thus preventing the escape of sap and the entry of rotting organisms. But in the first place there was no great loss of sap from the feeding punctures, apart from that taken by the aphids, and in the second chemical analysis did not show the Americans to have any more resin than the *viniferas*.

Foëx thought that the difference in behaviour was due to the fact that the insect puncture led to a flow of nitrogenous material to this point, the conversion of starch to glucose and the formation of an acidic substance leading to hypertrophy and death of the tissues there. These substances spread widely in *vinifera*, whereas in species such as *æstivalis*, *riparia* and *rupestris* the toxin did not move, due to the formation of scar tissue blocking its progress. Such tissues, he held, eventually sloughed off. As a consequence Foëx thought the prospects for the success of grafting were bright. Recent research (*see* page 184) suggests that the resistant Americans possibly protect themselves by some such mechanism, as do certain other plants against attack by disease-causing fungi: in such cases blocking tissue can form, stopping the spread of infection. Such an ability is frequently a basis for resistance to disease in certain varieties of plants.

Foëx was not quite right, but was moving in the right direction; the truth seems to be that the aphid injects a hormone-like substance into the plant, a chemical far more active in *vinifera* than in the resistant American species; we will discuss its nature later. In the 1870s M. Boutin (Ainé) analysed healthy and attacked plants and tried to find explanations for the differences between them.[27] The healthy vine roots had more sugar and starch, in fact more of everything except residual ash, than the attacked roots. Ash was 12·85 per cent in attacked roots, and double that of the healthy roots.

The fact was that the pest was killing the roots and depriving the plant of water (*see* Note 14) and the power of making plant food. The aphids seem to have drawn all sugar from the bark area and one might also assume that the attacked plant has had to get food from where it could and was also drawing sugar from this zone. The healthy roots had far higher food reserves. The doubled ash content of the attacked rootlets compared with the healthy ones shows that the former were suffering from water shortage. There is not much difference in the ash content of the two kinds of leaves, suggesting that these had priority in drawing on scarce water supplies in the plant.

It was thus gradually accepted that the industry could be saved: two parties began to form; the 'chemists', believing in carbon bisulphide, sulphocarbonates and water, and the 'Americanists'. The turning point for the victory of the Americanist school was the International Phylloxera Congress at Bordeaux in 1881 at which one of its principal spokesmen was a remarkable woman, Marguerite-Auguste-Marie, Duchesse de Fitz-James, who had a vineyard at St Bénézet, Gard. We ought to know a little more about this lady. She was the daughter of Charles-Frédéric, Count Löwenhjelm, Swedish minister in Paris. She married Edouard-Antoine-Sidoine, Duc de Fitz-James (the sixth duke) on 17th May 1851,[8] and in due course had four children. The Fitz-James were descended from an illegitimate son of James II (of England) and Arabella Churchill, this son having been created a duke by Louis XIV, with a *seigneurie* at Wartz, Oise, later renamed Fitz-James and existing under that name to this day. The Didot-Bottin for 1878 shows the duke and duchess with separate addresses in Paris, so presumably they had parted company by this date, the duchess residing with her elder son, the Comte de Fitz-James, in a house no longer standing. Her two daughters were married by this time, leaving the duchess free to turn her mind to other matters, the cultivation of her garden (vineyard) being a principal activity. Presumably her native language was Swedish, she wrote well and effectively in French and one is entitled to think that she was also fluent in English as frequently she makes use of English expressions in English in her works ('*Les anglais disent* "If you want a thing doing do it yourself"' [73]). Obviously she had a tremendous grasp of the realities of the situation. Her title enabled her to catch public attention, and her logic carried conviction.*

The duchess's description of the Bordeaux Congress is both lively and rational.[72] The Congress was postponed from 29th August to 10th October which, the duchess maintains, greatly influenced the dispute between the rival factions. If the victory

* The Abbé G. Dervin made a tour of the southern vineyards in 1895 and describes the duchess as being small and most energetic. Living in the mountains, she drove through them every day in order to supervise her vineyards on the plain.[190]

of the American vines did not seem to be very impressive, this was due to the excess of eloquence towards the end of the Congress, leading to the postponement of the closing session from Thursday, 15th October, to Saturday, 17th October, when many 'Americanists' had returned home to draw, or sell, the wine they had in barrel whose quality, whose existence even, was being questioned. Of the two groups—'chemists' and 'Americanists'—she said 'these words advantageously may replace those actually used'. One would like to know what the actual words were.* 'Stinkers' seems possible as one of them and 'wood merchants' for the other.

The chemists were well organized, under control, always in attendance, expressed approval or disapproval as a block and were so disciplined that they gave the appearance of a first night theatrical production, a reference to the *claque*, so frequently found on such occasions. As to their position in the conference room—they made the duchess think of the military defensive square. On the other hand the Americanists were so sure of the truth of their cause that they were just scattered throughout the auditorium and came and went as they felt inclined. The Americanists had travelled a hard road. She (the duchess) had been a fervent 'chemist' to begin with, buying her way out of this path slowly, because everyone was fearful of facing the trouble, uncertainty and expense of replanting. The first man who came to sell her American plants received very rough treatment, not being allowed even to leave his name and address! How things had changed; now the chemists were in retreat. There was a third party, though, between these two whose characteristic was the denial of all hope, reminding one of the thousand-headed hydra—cut one off and another grows. *Savants* were considering the most childish objections, forgetting, whilst MM. Balbiani and Boiteau searched for the winter egg, that the years pass, that death strikes and hunger kills. Even if it were true, as they say, that American vines would last only fifteen years, it is better to plant them than remain inactive. Instead of being so pessimistic these people should take a trip through l'Hérault, Gard and Vaucluse to see the giant strides American vines had made. At Pignan they

* About 1958, when the Chemical and Authors' Clubs of London took joint premises to save money they were affectionately known as 'The stinkers and thinkers'.

would see Aramon on *riparia*, at St Bénézet Oeillades on Taylor, and Jacquez at Décapris, Var, all flourishing.

'On seeing these beautiful vineyards,' said the duchess, 'one would have to be a "*systematic incredulist*" not to believe in them. We are at the Holy Saturday of viticulture, on the eve of its resurrection.' She opposed M. de Laroque's views and several agricultural papers supported her, and she thought that cheap-rate excursions through these 'Americanized' vineyards should be arranged so that growers could see the results with their own eyes. It was Planchon's visit to America which had opened the way to success for both the French and the Americans, particularly for the Californians, whose vineyards were recently attacked by the pest and who now saw that grafting, as starting to be used in France, was the remedy. It was in France, not in America, that the answer was to be found.

The duchess did not mince her words about M. de Laroque, 'that pall-bearer, fighting the phylloxera on the dead body of the victim itself'. He was both warrior and diplomat, well knowing that words were given to man to hide his thoughts, or rather his fears. . . . If he was so right why had the masses not rallied to his banner . . . ? However explosive and latent his thunder, called carbon bisulphide,* it has blown up neither the Americanists nor the phylloxera, but is said (and there are bad enough people around to say such things) to have killed two hectares of a famous château. The meeting had patiently listened to M. Fallières on the advantages of CS_2 and sulphocarbonates and had coldly concluded that one could hold the pest in check more or less successfully with such substances, much trouble and money. M. Gachasin-Lafite's solution (grafting) was much more attractive, bringing the vines back to normal luxuriance, to the Garden of Eden even. It would be better to see what M. Gachasin-Lafite had seen before bestowing the epithet 'wood-merchant' on the Americanists. Why was this word pejorative? What was wrong with selling one's surplus American wood and enabling others to take advantage of this method? Could M. Fallières tell her of any Americanist who had turned to insecticides? She could name many users of insecticides who had turned to American vines!

* This suggests that CS_2 explosions were commoner than appears to be the case from contemporary records.

She did not maintain that insecticides were ineffective and had in fact treated 13 ha. twice a year from 1872 to 1876. It was difficult to use insecticides on a large scale, difficult to find skilled labour at any price; one's crop was at the mercy of a moment's forgetfulness, or of some crisis such as a war or a death in the family, and she had abandoned the system in favour of American roots.

The price of American wood depended on supply and demand, as with any other new plant, such as new roses or camellias. It was absurd for Dr Despétis to maintain that Americanists were only in it to sell wood and could make 50 fr. a vine at it.

Flooding would kill the pest and was suitable where it could be done cheaply. It could not be done on hillsides, nor could insecticides be used there, because of the expense. To advocate the use of insecticides there was like telling a sick, poor man to spend his winters in Madeira—impossible.

The duchess praised MM. Planchon, Laliman and Douysset. The first for disposing of the rumours about the superiority of Sudan vines and the two latter for their work on anthracnose and mildew. The duchess ended by thanking the Congress president, M. Lalande, for his impartiality and conduct of the meetings. She had spoken with some trepidation, she for had read somewhere '*Taceat mulier*'.*

She had some very practical ideas on spreading information and had realized that this was the stumbling block. In a letter to the President of the Bordeaux Congress she proposed that the railway companies lower their tariffs for the transport of plants and bud wood (there is nothing unusual about this; the lowering of transport charges is a constant theme among farmers) and that (and this was the novelty) every railway station in the vineyard areas should have five American plants (ones suited to its geographical position) growing around it, so that growers could see how they throve alongside the dead *viniferas*.

M. Prosper de Lafitte seems to have been somewhat of a bore and a fanatic on the winter-egg theory. In a later publication [73] the duchess describes an incident at the Lyons Congress of 1880. The hall for the meeting (or possibly tent) had been hired from the travelling Pezon circus and menagerie. Prosper de Lafitte

* Let a woman be silent.

had not been received with much cordiality as he was so pessimistic and would not stop talking, and as time was up the proprietor wanted to get the conference out and start his evening performance. The crowd were whistling and impatient to get in. M. Lafitte continued his discourse and the circus owner let out the monkeys. 'The strangest scene then followed,' says the duchess, 'the audience laughed, the proprietor raged, the would-be new audience stamped their feet and shouted. The monkeys, astonished or amused—I do not know which—surrounded M. Lafitte and made faces at him. He, firm and noble to the last, continued to speak, defying alike both man and monkey!'

Another lady, Mme Ponsot of Pomerol (she might be described as the Fitz-James of the Gironde; the Société nationale d'Encouragement à l'Agriculture gave her a gold medal in recognition of her mildew work) also helped in the victory of the Americanists and a standing ovation was accorded both ladies in one of the final sessions of the 1881 Congress.[24] An additional factor was an exhibition of wines from American species and from French varieties grafted onto American roots. Fortunately for the cause the tasting jury, led by M. P.-A. Labrune, came to the conclusion that they could not judge these wines properly because they had been made in too small quantities, drawn too soon in order to be ready for the Congress, tasted too soon (it was the middle of October) as well as being altered by the journey. The jury made no awards to the wines, but gave a gold medal to M. Piola, of St Emilion, for his exhibition of such a wide range of wines. (It could then just be claimed that wine from American vines had received a gold medal.) The Congress had to make its judgment on this matter from some analyses presented by M. Bisseuil, from Charente-Inférièure, for Professor Xambu.

The professor concluded that the famous Cognac grape, the *Folle Blanche*, gave the same wine when grafted as when on its own roots, and the same remarks applied to Quercy and Aramon. Of the Americans, Jacquez and Norton's Virginia gave highly coloured wines (Note 16). Herbemont, Cunningham and Elvira gave brandies very different from the true cognac, and the other Americans might do for table grapes.

The logic of the situation gradually imposed itself even though some of the grafted plants failed, due to using but slightly resistant American species (*labrusca*), or because Americans liked acid soils and many of the vineyards had chalky ones. The first area to take to Americans was the Midi, followed by the Gironde; Burgundy resisted for a long time, even prohibiting the import and planting of American varieties from 1874 until 15th June 1887.[110] The proprietors of the famous vineyards were afraid of spoiling the reputation of their *grands crûs*. Needless to say there was much clandestine introduction of Americans, and in any case the prohibition did nothing to stop the spread of the pest constantly being borne in by the zephyr. These illegal immigrants can be found in remote places to this day. In far-off hill villages a peasant proprietor may offer you a glass of his '*vin du pays*' which is execrable; this is surprising, because such wines, made in small quantities with loving care for home use, can so often be very good. Before you blame bad vinification, dirty barrels and so on have a look at the vine itself. Quite likely it will be a *riparia*, *rupestris* or even a *labrusca*, planted in despair in the 1880s and still living or kept going by *provinage* (layering of summer shoots) from time to time, thus still cropping and giving an American vintage of foxy-tasting wine. They need not, of course, be pirate vines but ones planted after permission was given to do so.

The task of 'reconstituting' nearly all the French vineyards was an enormous one and a simple calculation shows this. In France in 1880 there were 2,209,000 ha. planted to vines. A modest estimate of plants is 5,000 per hectare; there were thus some 11,045 million vines growing in France. To make a new plant some 25 to 35 cm. of ripened wood is needed for the rootstock and a few centimetres of wood (one bud) for the scion, say 30 cm. of good wood per plant: 11,045 million × 30 cm. = 3·3 million km. (2 million miles) of ripened wood 8 to 10 mm. thick. Unripened and thin wood to the same amount would be discarded, giving us, say, 7 million km. of wood needed; 7 million km. is a big figure: it is about ten times the distance of the moon from the earth. The weight of wood needed is also quite impressive. A cutting weighs about 20 g. and not all cuttings 'take', so an excess is needed. Even assuming a 100 per cent success the 11,045 million cuttings needed in France alone would weigh about 220,900 tons. The actual wood used would be about double this figure. There was some justification in calling the new nurserymen that arose to supply these plants 'wood-merchants'. (Both Italy and Spain, with larger vine areas, would need similar quantities too.) At that time (1881–2) cuttings of American wood were on offer, bare, at 80 fr. to 125 fr. the 1,000, and rooted at from 180 to 250 fr. the 1,000. The potential French market was thus between 884 million and 2,760 million fr. (£35 to £110 million). French *vignerons* did not pay as much as this though, for growers started making their own plants, using simple grafting machines.

The first obstacle to the success of 'Americans' was overcome by creating a list of relative resistance to the pest. Millardet established a scale running from 0 (no resistance to the root form) to 20 (complete resistance to the root form). (*See* Table 7, page 212.) MM. P. Viala and Ravaz [149] also made a resistance scale with the same maximum and a slightly different distribution of the species. In their table *rotundifolia* received 19 marks and *rupestris* 18. No doubt soil differences accounted for the variation which, after all, was not very great. In 1966 M. Boubals and a colleague, M. Pistre,[25] published a study (*see* Note 17) establishing two new scales for phylloxera resistance, running in the opposite direction to the older ones, from 0 (complete resistance) to 3 (susceptible) for greenhouse tests and from 0 (very resistant) to 5 (very susceptible) for field tests.

The second obstacle to American stocks, that most of them did not like acid soil, has already been mentioned; this led to a great deal of disappointment and complaint. Vines affected in this way go chlorotic, and today we know that more often than not they are suffering from lack of magnesium, the carbonate in some way preventing the plants' uptake of this essential element. Viala did much to dispose of this problem. In 1888 he visited America, listed 430 varieties of American vines, analysed soils and came up with a list of calcium tolerances to aid the French growers.[179] Briefly, his findings were that all American species would grow with less than 10 per cent of calcium carbonate ($CaCO_3$) and that the only species that would grow in soils with more than 60 per cent $CaCO_3$ was *berlandieri*. He also fixed intermediate carbonate levels for other species and varieties which, in general, hold to this day.

An important point is the *distribution* of the chalk in the soil; if it is in comparatively large lumps or nodules a susceptible vine will tolerate a higher percentage than if it is evenly distributed throughout the whole mass.[40]

It soon became evident that better stocks were needed and could be obtained by breeding, which was at once put in hand. Obviously there was going to be an immense demand, and nurserymen saw a very profitable future in this business—one far better than roses and camellias. We may note, in passing, that a hybrid called Vialla was bred by MM. Viala and Ravaz, named, presumably, in honour of their entomological colleague with the double 'l'. It was not of much practical use, though many others were, and the production of new stocks continues all the time.

There were a number of Cassandras who predicted disaster for the industry, and even as late as 1908 Professor L. Daniel (of Rennes University) wrote an article for *The Times* [174q] denouncing the practice. He maintained that the grafted vine produced a vast quantity of grapes at the expense of quality. The growers had found in grafting 'a Pactolus * and after Pactolus came distress and ruin'.

In one respect M. Daniel was right. The new plantings enabled vineyards to be laid out more satisfactorily and yields to be increased, but it was not the American stocks that were increasing production but the planting on richer soils, more manuring and better general attention.

Some American species (such as *riparia*) were difficult to root, and to overcome this they were grafted onto *vinifera* in phylloxera-free areas with the graft union and several buds above it buried in the soil. This helped the American vines to root much more rapidly.[17]

Grafting plants had been an accomplishment of French nurserymen for a long time and they knew a great deal about compatibilities of stocks and scions, that only plants of the same or of closely related species would 'take'. In contrast, much of the general public had but the vaguest idea as to the possible, and we have already mentioned the suggestions of grafting onto blackberries, mulberries and other plants by the would-be prize-winners—all quite impossible. Bazille, Laliman and others showed that *vinifera* and the American species were compatible, but the success rate was greater between certain pairs than with others. The question also arose as to what kind of graft to use (the type of joint to be

* Pactolus. The Lydian river, the sands of which supplied the classical Grecian kings with gold, the source of their wealth. The gold was placer and its surface supply was soon exhausted.

made) to secure best results, particularly when it was realized that millions of plants would be needed.

Here we need to consider the theory of the matter for a moment. A woody stem roughly consists of a central woody core, surrounded by a green or greenish cambium, the vital growing area, and that surrounded by a layer of bark. The shoot expands by the cambium laying down more wood on its inside and more bark on its outside and by expanding itself. Since it is only the cambium that grows, the two cambiums must be brought together under such conditions as will encourage the cells of stock and scion to grow, mingle and unite into a strong whole. Consequently both parts need to be in good condition with good food reserves, and as much of the two cambiums should be put into contact as is possible; obviously in this way there is a greater chance of the joint being made and it will be all the stronger the larger the joining area is. There are many kinds of graft (plant grafting, that is). The most popular in France became the whip and tongue graft, known there for some reason as the *greffe anglaise*, presumably having been introduced from England. From Fig. 9 it will at once be seen that the tongue and its corresponding slot make for contact of a much greater cambium area. The whip and tongue graft needs stock and scion to be of about the same section area, and soon diameter gauges of small size were available for this purpose. Graftwood was always sold as 'ripened', meaning it was autumn cut with a good food reserve and of a certain length and diameter. The whip and tongue must fit well; thus the slope of the cuts on stock and scion must be the same, and the slot (whip) must be just deep enough to take the tongue. French ingenuity was soon at work on this problem and grafting machines came onto the market, much to the scorn of the *maître greffeur* who, with his razor sharp knife, unerring cut, joined, tied with raffia and waxed his wood as if by magic. Grafting schools were established (*see* Chapter Twelve) and a whole new skill was born. It was eventually found to be more satisfactory to make the graft before getting roots onto the stock. If a rooted stock was grafted and did not 'take', the stock was usually wasted; it was the stockwood that was rare and expensive. This was known as bench grafting and was where the machine could be used.

The work was done (and is done to this day) indoors, usually in a large shed, and the work is systematized. One gang cuts the

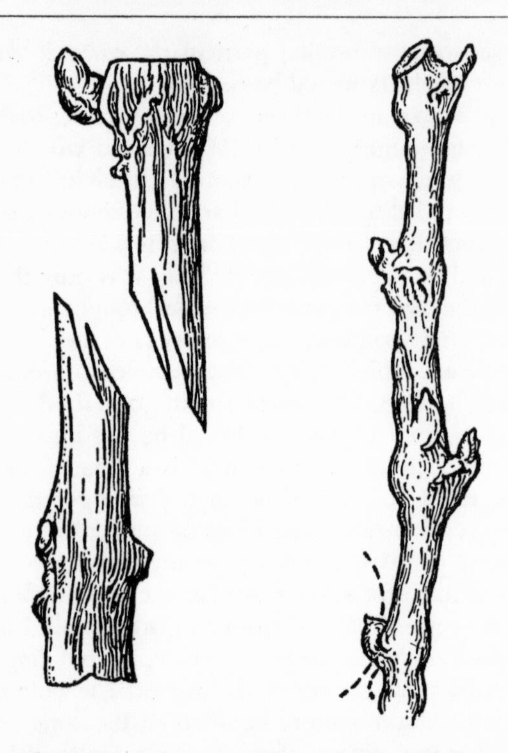

9. *Whip and tongue graft, or* greffe anglaise. *Grafts were made in millions and so skilled did the operators become that no tying was needed. The scion above is a* vinifera *and the stock below is an American species. The bud on the stock is rubbed off to encourage rooting.*

scions, each having one or two eyes (buds). Another gang prepares the American wood, all the eyes except the top one being removed. A third set of men (or women) cut and make the grafts which are then packed into boxes and surrounded with a mixture of moss, seagrass and sawdust (growth hormones are sometimes added today). The boxes are wetted and kept in a hot room, 25–30° C. (the callusing room) for twenty days. Here a jelly-like growth (the callus) appears around the joint. From the hot room the boxes are moved to the hardening room at 15–16° C. for a few days. They are then inspected, and the sound joins retained; usually the joint is strong and will withstand a 5-kg. pull. Often roots have started to form. The joined wood is now placed into trenches in a nursery bed to complete the rooting. Obviously any roots which may have formed on the scion are rubbed off. Under these conditions the whip and tongue grafts are so firmly and precisely made that they do not need tying or waxing.

CHAPTER TWELVE

Social and Economic Effects

I Château Loudenne, Côte d'Or, Loire and Champagne

The archivist of the city of Hereford, Miss E. M. Jancey, in a recent address to the British Agricultural History Society said that, apart from an extensive eruption of volcanic ash, the next best thing, from the historian's point of view, is an absentee landlord, always provided the correspondence survives; usually immense and interesting detail is documented. Such is the case with Messrs Gilbey of London, who have kindly allowed their Bordeaux archives to be inspected, enabling us to uncover in this way much of interest in the phylloxera story. Other documents have also been used.

The Château Loudenne and its vineyards, in the Médoc, Gironde, were purchased by Messrs W. and A. Gilbey in 1875 from Madame la Vicomtesse de Marcellus for 700,000 fr. It was an unfortunate moment because they stepped straight into phylloxera trouble. There were 190 ha. of land, of which 50 were in vines. The new owners spent one and a half million fr. on constructing a quay onto the estuary, many buildings and extensive cellars which would hold 16,000 barrels and half a million bottles. Though Loudenne was not a 'classified growth' it had a high reputation for good wine. The local villagers were much impressed with the 'enormous English horses' imported. In spite of their difficulties the château obtained the gold medal for wine in 1887 and by 1892 had 100 ha. in vines, all new plantings, Sauvignon and Cabernet on *riparia* and *solonis* stocks. The path to this happy conclusion is described below and must be similar to the thinking and action of thousands of wine-growers throughout Europe.

Correspondence between Mr E. Brown, the manager at Loudenne, and Messrs W. and A. Gilbey in London was conducted in French by Mr Brown and in English by Messrs Gilbey and seemed to work very well that way. In a letter dated 1st July 1878 Mr Brown complains of the terrible wet weather, which he fears will bring the *coulure*, though there is very little at Loudenne at the moment (*see* Note 7). What he fears most is the phylloxera, which is worrying people in the commune of St Yzans and which appeared in the Bas Médoc the previous year. The pest had, in fact, appeared in the white vines, No. 10 plot, at Loudenne, though it did not seem to have spread.

Gilbeys replied on 9th July that they too were worried and that Mr Brown should take the active steps against it they had already discussed. These seem to have been to rip out the white vines and replace them with reds, which, of course, as they also were *viniferas*, would also have been attacked and yet more severely, being young.

Mr Brown realized this and said so in a letter dated 13th July, quoting many cases where the pest had destroyed reds (in passing one may note how quickly letters travelled in those days. Could such speed from a remote commune be maintained today? One doubts it). There was no effective remedy for the pest. He proposed to pick what grapes he could and then to pull out the vines, replant the area with cereals or forage and not replant with vines until the remedy for the attack had been found. Taking this crop may well have been a mistake, because as soon as the weather cleared and the soil dried and cracked the insects would have swarmed out of the ground and have spread the infection.

Gilbeys agreed to this course, and on 20th January 1882 we find Mr Brown negotiating with the Société nationale contre le Phylloxera to treat the Loudenne vines with potassium sulphocarbonate. But, alas, the Société cannot obtain supplies of the chemical and cannot promise treatment until April, whereas Mr Brown wants the work done in March, a situation familiar to spray contractors to this day. Mr Brown also suggests applying for the admission of Loudenne to the newly formed vine-growers *syndicat*. On 22nd January (note that the letter took but two days) Gilbeys replied agreeing to join the syndicate.

In March Mr Brown has joined this body and says the spray contractors still cannot get sufficient sulphocarbonate. There is only one factory in Paris making the stuff and they cannot produce enough. It costs 60 to 65 fr. the 100 kg. and is packed in barrels of 250 to 260 kg. Can any be obtained in England? Gilbeys seem to have had some difficulty in identifying this chemical and write asking for a sample to submit to analysis, to which Brown replies (19th March 1882) that he cannot supply one because it is a liquid, which is surprising in its way. As their whole business was the supplying of liquids, what was so difficult in the sending of a sample of sulphocarbonate solution? The fact was that the Société nationale contre le Phylloxera now looked like starting work. When they had this letter Gilbeys replied (21st March 1882), that they had had two quotations for sulphocarbonate of £28 and £26 per ton, f.o.b. London, presumably having found out what the chemical was in the interval and doing without the sample and analysis.

Mr Brown (25th March 1882) is pained that the price is so high in England (the price f.o.b. was about the same, but freight, etc., would have had to be added to it), but even with the shortage apparently is not tempted to procure supplies from London. Gilbeys next suggest (4th April 1883) that Loudenne undertake anti-phylloxera work on their own account rather than depend on contractors, and Brown decided to do so and to use the Société as well. Gilbeys were optimistic at this point and proposed to put all available land into vines. Alas, on 1st July 1882 Brown reports that the pest is spreading and that, though the treated vines look well, he doubts if ever the phylloxera will be destroyed. In other words the first phase was over—that the pest might be wiped out—and now they were at this second—living with it.

The insect continued to spread and the price of wine to rise. In 1883 they found out that two treatments a year were necessary (1st December 1883). By January 1884 they were still uncertain about American rootstocks, though as far back as 1882 (Gilbey's diary entry 23rd September 1882) American vines had 'impressed' Mr Gilbey, who wrote of the damage done by anthracnose, mildew and phylloxera and seemed to think the first trouble was the worst (which is doubtful). Because of the shortage prices were high—

> ... all absurdly high compared with the cheaper Haut Médoc wines, being for 1879's 1,000 fr., for 1880's 800–950 fr., for 1881's 950 to 1,150 fr., and in one or two instances they are holding out for 1,400 fr. for 1881's.* We should have returned very discouraged except for the fact that on visiting Laujac we were impressed by the means which had been taken there to meet all these evils, and the quantity of American vines being planted served to confirm the idea that this method will probably be adopted as a means of renewing and reinstating the Bas Médoc vineyards.

He found the Loudenne vines looking very well as they had been treated with sulphocarbonate. The 'Americans' he saw at Laujac were no doubt American species, not French varieties on American roots, and Mr Gilbey had yet to taste the wine made from them.

In 1883 Gilbeys started to replant on American roots with some fears, still present in January 1884, but by the autumn of that year they were convinced that grafting was the obvious answer. The Midi and the west took to it far more readily than did Burgundy, whose wine story has been very well described by M. Robert Laurent,[110] to whom I am indebted for much of the following.

In Burgundy a common attitude to this threat seemed more prevalent than in the Midi and the Gironde, though it was strong enough throughout all France; it may be described as the 'it can't happen to me' syndrome. It is widely found to this day; *my* worn tyre will not burst, smoking will not give *me* lung cancer. But, just the same, it happens. Firstly, the pest is much talked about—a nasty little insect—then a neighbour discovers it; then one gets it oneself and rushes to the chemist to buy a remedy, as fate removes more and more vines every day and ruin and disaster spread. Even the experts are denigrated. Planchon, Bazille and Sahut, who first found the cause, were mockingly called 'the Hérault entomologists', 'because not one of them is an entomologist by profession and the number of ignoramuses who think they know more defies arithmetic'.[71]

On the Côte d'Or the pest was first found officially in Mersault and Dijon in July 1878 when the Prefect sent out a notice about it. The authorities had been slow to react, for in

* Per *tonneau* of 900 l.

July 1871 the Minister of Agriculture had sent out a circular advising local authorities to form study and vigilance commissions; it took Dijon three years to do this, even though the pest was quite near, at Mancey, Saône-et-Loire. This commission, formed 30th July 1874, passed a decree in September of the same year prohibiting the import of vines of all sorts to the area; obviously they hoped to be safe from the pest in this way. By 1875 the danger was growing and a sum was voted for the work of the phylloxera commission—200 fr.! Not a large amount. Even so, and though raised to 600 fr. for 1876 and 1877, it was never all used, pointing to a general disbelief in the phylloxera threat, in fact the local government seemed to be wilfully blind. In August 1876 they said: 'The pest's advance has slowed down leading us to hope we shall be spared.' As early as 1874 they had refused to vote a 10,000 fr. prize for a phylloxera cure, which possibly was just as well, as such prizes rarely, if ever, achieve their object; nevertheless this displayed their mental attitude to the menace.

Other people were not as inactive as the area commission. In Beaune in 1877 a municipal commission was formed and at Nuits a public subscription raised 1,200 fr. for phylloxera studies, people being sent to Montpellier to learn what was happening and what could be done. This work led to a M. Viard, of Mersault, recognizing the presence of the pest on his vines in July 1878; a few days later it was found at the Dijon botanic gardens and in September at Norges-la-Ville. In spite of this the local government were still optimistic and clung to the phylloxera/effect theory. The insect, they said, only got onto unhealthy vines, such as those pushed to give 300 hl. of wine per hectare. The carefully grown Côte d'Or wines with moderate yields would not be attacked. As to the peasants and smallholders, the old hands were believers in 'elbow grease', plenty of hoeing would destroy the pest, an action actually more likely to lead to its spread.

The authorities' immediate reaction to the discovery was to destroy infection centres and prevent re-import of the pest. Soldiers were posted around M. Viard's vineyard to prevent access and the P.L.M. railway asked to treat the area with carbon bisulphide. The company could not undertake the task and, local labour being unwilling, the soldiers did the work. It was then announced that all infection centres had been destroyed, a cause for rejoicing. An examination of the treated centres, however, had shown that the infection was some three or four years old and that there must have been many other infected vineyards in the area. As a consequence, in October 1878, the mayors of communes between Dijon and Beaune were urged to organize a phylloxera search, but no systematic examination was made because 'the phylloxera announced itself'; true enough, but by the time it did this it had released swarms of infecting insects to spread the trouble.

By August 1879 thirty-four infected centres were noted in Beaune: the pest was becoming general and more active measures needed to be taken. By this time the central government in Paris had passed laws for phylloxera control (law of 15th July 1878) and Burgundy was fortunate in one respect. The 15th July law said that in newly infected areas treatment would be paid for by the state. When the law was passed neither Dijon or Beaune were healthy, but the minister did allow the new law to apply to them.

Growers on the whole did not believe the evidence of their own eyes. The pest had existed for all time . . . it was just the result of no proper care of the vineyards. The first CS_2 treatments killed both insects and vines and the news that compulsory searches and treatments were to be made got a poor reception. Political capital was made out of this general discontent. The Bonapartist paper *Bien Public* encouraged its readers to resist the commission, saying that the treatments were a plot to destroy the common wines for the benefit of luxury wine-growers. Some villages were divided on the issue. At Bouez the mayor's party was for treatment, destructive if need be, but the opposition—no treatment here—was much more numerous. The opposite was found at Puligny, where the party leaders were in favour of treatment.

What great quarrels and discussions there must have been, even riots. In July 1879, at Bouze, the demonstrators were not quietened by the presence of the Prefect in person and the *gendarmerie* had to be called out to restore order. Even so, the work had to be stopped. At Chenôve 160 growers chased the treatment team out of the area, saying they were *canaille*, more to be feared than the phylloxera itself, and a proprietor, presumably the ringleader, having been fined for refusing to admit the team, was acquitted by the Dijon court on appeal, an event soundly cheered and one encouraging resistance to treatment among the remaining growers.

Naturally the phylloxera continued to spread.

Lawyers found out that only in healthy areas could compulsory searches and treatments be made; of course it was as, or more, important to destroy the pest in neighbouring infected places. New legislation (law of 2nd August 1879) stopped this loophole and led to more trouble in Burgundy. The Prefect at first got tough and fined a number of growers at Bouze (11th October 1879) and then turned to conciliation. He began to use local labour, hid his soldiers, paid for any crop destroyed and dismissed unpopular officials; only gradually were the control measures accepted, mostly one imagines because it could be seen every day that the pest really was a killer.

By 1881 the Government gave up compulsory treatments, perhaps the last straw being a claim by a vine-grower for the sum of about 78,000 fr. compensation for 19 ha. of vines treated. It was now up to growers to defend themselves. Infection centres had run together and phase 1 was over. The phylloxera could not be eliminated and was becoming generalized.

The local government still interested itself in the subject and would treat vines where requested. For administrative and compensation purposes they voted 20,000 fr. a year from 1882 and such sums were matched, franc for franc, by the grants from the Central Government in Paris. Attempts were made to form phylloxera *syndicats* without much success, and a general inertia took over as the pest spread. The budgets even were not all spent. The pest got worse and growers eventually woke up to the facts of life with the insect. By 1886 there were sixty-six *syndicats* in the area and treatment was found to cost 250 fr. per hectare, a high figure, supportable for the noble vines, but representing 12 per cent of the revenue of ordinary vineyards at prices current 1880–1889.

The government subsidy for treatment also dropped, from 80 fr. per hectare in 1882 to 20 fr. in 1891, and no syndicate could claim for more than 5 ha. treated. After an existence of five years all subsidy ceased. It was only the famous *domaines* that could afford the expense. Lesser growers did what they could; abandoned the Gamay variety and kept the Pinot, which seemed to show some resistance, started to think about American vines (phase 3) as did the rest of France. As in the Midi and Gironde two schools of thought began to form; 'Chemists' (or 'sulphurers'), and 'Americanists', the big and the small growers.

The former, jealous of the good name of the famous wines (Aloxe-Corton, Pommard, etc.), wanted compulsory annual treatment (many treated twice a year), failing which the vineyards should be destroyed, and prohibition of all foreign vines in the area. The small growers, defeated, saw hope only in the American varieties, either as direct producers or as stocks.

The introduction of American vines to Burgundy had been prohibited by a decree of 1874. Action to repeal this edict grew and the first main mover was the Société vigneronne de Beaune, founded in 1885. The Burgundy Vigilance Committee was against the proposal and 'saw salvation only in insecticides', as the society commented in 1893. The Beaune Society had to feel its way very carefully and had the intelligence to pursue a mid-course. Its secretary (M. L. Latour) made a visit to the 'reconstituted' vineyards of Beaujolais, seeing experimental plantations and resistant varieties. He then proclaimed the need to keep the fine old Burgundian vines growing as before by the use of insecticides and at the same time planting out destroyed areas with the resistants. Moreover he printed 800 copies of his report on the above lines and sent them to senators, members of parliament, local councillors and so forth. The press got to hear of the matter and joined in with attacks on 'the supposed anti-phylloxera committees who live from our disasters' and 'cheap wine merchants making fortunes out of raisin wines' (*see* page 148), neither of whom wanted to see the problem solved. The public warmed to the question: if the 'bourgeois' would not let the small proprietors replant Americans it was to get cheap labour for their own, sulphured vines.

The struggle was long; a decree dated 15th June 1887 allowed the introduction of foreign vines into the Côte d'Or. Of course they had been smuggled in for some time, to such an extent, in fact, that the local government renewed the prohibition decree in 1885. In 1886 the annual report sent to the main phylloxera commission in Paris admitted there were at least 100 ha. of American vines in the area. Their obvious success, at least as far as growth went, must have led to the 15th June decree.

Many problems remained to be solved. Firstly, direct producers or grafts? Secondly, what varieties suited the soils? Thirdly, whence did one obtain the vast quantity of plants needed? The Beaune Society played a big part in overcoming these matters. They turned down direct producers—the wine was (then) so poor. The great thing was to spread a knowledge of grafting among the growers, to keep down the expense of obtaining the new plants. The society brought in three skilled craftsman grafters from Beaujolais in 1886 and the following year set up free grafting courses in thirteen *communes*. The more skilled pupils become instructors in their turn, spreading the good news. The enthusiasm for grafting can be seen in the figures of attendance at the courses, the number of diplomas awarded to successful candidates at the examinations and the fact that the diploma was usually framed and hung on a wall in a man's or his parents' house in a place of honour next to his certificate of military service.

In 1887 (the second year), forty-eight pupils gained twenty diplomas (1 per 2·4 pupils). The standard of efficiency demanded must have been high, but in 1888 only fifty diplomas were given among 1,000 pupils (1 per 20), and in the following year seventy-four among 1,200 pupils (1 per 17·5).

After 1900 there was not much need for such courses, for a knowledge of grafting became an accepted accomplishment for any aspiring wine-grower. The technique was passed from father to son. For instance, in another area, as late as 1909, M. F. Richter, of Montpellier, made seven million grafts a year and employed over sixty women in his grafting shed. Each one made about 1,500 grafts a day.[138]

The Beaune example was followed by other wine societies in the area. By 1890 it was no longer a question of whether one should or should not graft, but only one of what sort of graft and what sort of stock to use. Here again the Beaune Society took the lead. In 1887 they organized a nursery and in 1888 established a 'graft competition'. Obviously they could not then judge the success of grafts planted that year and they allowed owners who had 'defied the rigours of the law' by illegally planting grafted vines before 1887 to enter, mute evidence of widespread illegal imports to the area.

Having lost its battle with the Americanists, the Vigilance Committee, with commendable rectitude, encouraged grafting;

in other words they began to believe the evidence of their own eyes and pockets. Inspired by the Beaune Society they established nurseries for American stocks at Dijon and Beaune. Success was not universal, because little was known about the suitability of stocks and soils and about what came to be called 'adaptation' to soils. On the surface this suggests a Lysenkoist belief that any given cultivar through successive vegetative propagation gradually 'adapted' itself to local conditions and became more suitable which, as we know today, is nonsense. In the absence of a bud sport—an extremely rare event—a *riparia* needing certain conditions (say less than 15 per cent of chalk in the soil), and planted today will still need the same conditions after any number of subsequent vegetative propagations. 'Adaptation' was constantly discussed and in nurseries it even appeared that American plants were adapting themselves to French conditions, so that believers in this theory appeared to be justified. What in fact was happening was that much of the American wood obtained was a very mixed lot, and that in the nurseries and *champs d'adaptation* the unsuitable strains were dying off and the more suitable ones were surviving. Even within a species there would be differing races, some more suitable for given conditions than others, so that in the popular sense 'adaptation' was taking place whereas, scientifically, only selection was happening.

The enthusiasm for the new viticulture was mostly among small growers, and the actual areas planted to grafted vines was small, 60 ha. in 1888 and 100 ha. in 1889. Fraud and ignorance held up the 'reconstitution' progress, particularly ignorance of the chalk content of soils, estimating which was a long and costly process until the *calcimetre* was invented in 1892, an apparatus capable of being used by any local pharmacist, or schoolmaster. A typical *calcimetre* was the 'Bernard'.[39] The method was essentially one of measuring the CO_2 gas given off from a gramme sample of soil when treated with hydrochloric acid. At the same time more experimental nurseries were being established by the Vigilance Committee. By 1891 there were 100 'adaptation nurseries' in Beaune and 44 in Dijon, and plants were sent out to 677 wine growers. By 1890 the area planted with American rootstocks was established at 1,000 ha. The amount of subsidy given to direct anti-phylloxera measures was reduced and a bigger proportion of it given to 'reconstitution.' The subsidy for 'reconstitution' was never very great, being 12 fr. per *ouvrée** (28 fr. per ha.) in 1892 and reaching its maximum of 17·12 fr. (40 fr. per ha.) in 1893 and 1894. According to Laurent this was only 8 per cent of the cost if the grower bought plants and 18 per cent of it if he made his own. It was not very much for a man who saw his vineyard and livelihood dying before his eyes. Even the Vigilance Committee described its aid as 'illusory'; growers had to make sacrifices, draw on their reserves or go out of business. The large growers were able to undertake the 'reconstitution' work by obtaining loans or drawing on these reserves; for instance the Hospice de Nuits made a loss of 7,220 fr. in 1894 because of the 'almost total destruction of its vines', whilst at the same time having to find considerable sums for the replanting. They were allowed to draw 6,589·27 fr. from contingency funds.

The state of mind of the vine-growers was terrible in those years as the vines died or were treated at tremendous cost. The small grower was not a farmer and rather despised that race, but he was being forced to join them if he did not emigrate to Algeria or the U.S.A. Those who stayed pulled out the dead vines and put in wheat and forage crops, cursing the Americans and their *mauvaises bêtes*; they also worked part-time, or even full-time, on the big vineyards, helping to apply the insecticides or flood water. In this way the feeling was fed that the famous vineyards opposed the introduction of American roots, because the misery of the small growers provided them with skilled, cheap labour, as no doubt it did (*see* page 160). But the pest was not actually *desired* by the big men. By 1892 things seemed to be getting better; the *mauvaise bête* had come from America and the cure for it was coming from that distant land as well. That year 420 ha. were replanted on American roots and 630 ha. in 1893. The Beaune Viticulture Society joyfully reported two great happenings of 1893 'in the vineyards one could see two things, vines again in the spring and, in autumn, grape pickers'.

This was the critical year. What would the wine from these grafted vines, many now three years old, be like? With relief it was found that it was indistinguishable from the old wines from the old own-root vineyards, though how much the wish was father to the thought is difficult to say. In March 1894 a gold

* There appear to be 2·33 *ouvrées* per ha.

medal was awarded to a Mersault *Goutte d'Or* at a Paris wine fair, and the Minister of Agriculture came in person to Mersault to present to the owner, M. J. B. Tavernier, both the medal and the cross of the *mérite agricole*. This must have been a tremendous occasion, the recognition that the vineyard would live again, that America was not wholly bad, that the sacrifices and misery at last were ending.

The local authorities still continued to subsidize insecticide treatments, the State, as before, making available a sum equal to that voted locally, but the proportion spent on defence gradually fell. The State also gave subsidies to the vine syndicates, which continued after the departmental subsidy was stopped in 1895. State subsidies were stopped in 1904 when they had been reduced to 10 fr. per hectare, a sum hardly worth the work of collecting.

In any case the total subsidies were not enormous; Laurent gives them as: 'Defence' (that is, insecticide treatment) 1882–1904, mostly carbon bisulphide, 947,000 fr.; American roots, 188,582 fr.

Difficulties with chalky soils began to arise, and some of the best slopes were still bare because *solonis* and even *riparia* would not grow in them. Nothing daunted, the now hopeful community established numerous additional experimental nurseries. In 1893 the Beaune nursery set up an annexe at Auxey, in a soil with 57 per cent of carbonate. Next year there were eight more in soils with 40 to 60 per cent of chalk. In 1895 a tenth was established on the Dijon slopes. In all these numerous, mainly hybrid, stocks, mostly Montpellier creations, but some from Mr Riley in America, were tested. There were fashions in stocks and many frauds too. *Solonis* was a favourite for a long time, but was abandoned in 1894 when the ever-active Beaune Society condemned it after a solemn inquiry,[165] recommending Couderc 3,309 and Millardet 101 in its place.

The progress of these events may be seen in Table 8, page 213. The figures are for a typical vineyard of 50 *ouvrées* (21·4 ha.); consequently, if doubled, they are percentages.

Table 8 clearly shows the struggle to keep the fine wines growing at the expense of the ordinary ones. In 1891 all the fine wines still alive (80 per cent of the original area) were treated, whereas in the same year only 37 per cent of the ordinary wines were—and the ordinary wines still alive were only 64 per cent of their original area. Replanting with vines grafted onto American roots went on at the same rate in both classes, with the result that the losses in the fine wines were less than in ordinary ones. Growers naturally would pay more attention to the better and more profitable vineyards.

The scarcity of wine did not lead to any great compensating increase of prices. Wines were imported in considerable volume, *piquettes* were made, and wines actually manufactured from imported dried raisins, about which we will say more later.

From 1892 the good news led to a spirit of optimism and a recovery in the sale price of vineyards. Between 1885 and 1891, when it looked as if the phylloxera would win the battle, vineyard values fell to half in some cases and to a quarter in others, but by 1896 they had regained the former values. The *grands crûs classées* were worth 15,000 to 30,000 fr. the hectare and good ordinaries from 8,000 to 12,000 fr.

Thus the Côte d'Or vineyard recovered due to the enthusiasm, application, intelligence, hard work and determination of thousands, but the vineyards there (some 23,000 ha. in 1878) lost about 6,000 people, mostly by emigration from the area.

The early part of the Loire story is put in here as an example of an intelligent and sincere approach to the problem. It is based mainly on the report of the Loiret Phylloxera Commission for 1877,[43] a clearly written document and one containing a certain amount of subtle wry comment, the author, presumably, being M. Maxime de la Rocheterie. The report summarizes the experiments made in 1877 in the Loire. It starts by regretting (but in very polite terms) that the members of the General Council who visited the vineyard area on 22nd August 1877 had but little time available for visiting the actual experimental plots (obviously they did not visit them at all but relied on the secretary's report) and it continues with a comment on the struggle for the prize:

> As with all pests which devastate humanity and menace the people's wealth, the appearance of the phylloxera ten years ago gave rise to an uncountable number of remedies, each invention being described by its discoverer, with complete confidence, as superior and infallible. The prospect of winning a prize of 300,000 fr. has once more both aroused covetousness and heated imaginations to such an extent that it is now almost impossible to make a complete list of all the processes put forward as cures.

The pest came fairly late to the Loire (1876), and inventors, repulsed elsewhere, fell on this new field with renewed avidity. The Commission tried first of all to exterminate the pest in the centres of infection and would have nothing to do with the 'inventors', but a number of wine-growers did try the various remedies proposed and the Commission comments on some twenty processes used privately—somewhat drily letting the results speak for themselves: '2. . . . four vines treated by M. Michel of Nantes, with a liquid of unknown composition: same result, the vines all dead.'

All the results were of this nature and frequently the disappointed owners called on the Commission for help. '5. M. Boison, Nièvre, treated eight lines of vines with an 'unknown liquid' with no result and seeing the difference between his vines and his neighbour's (which had been done with carbon bisulphide) asked the Commission to be kind enough to treat his eight rows when the time came to give the second treatment to his neighbour's vineyard.'

The Commission's own tests were done on a budget of 3,000 fr. (2,000 fr. from the General Council of the Loire and 1,000 fr. from the Ministry of Agriculture). One is always impressed with the vast quantity of work done on very small budgets. Six methods were tried, the Sabaté glove, pyrites from St-Bel, Boutin's mixture (largely sulphocarbonate and carbon bisulphide absorbed on chalk so that it could be applied as a dry powder), the Abbé Chevalier's rock (not a religious touchstone, but a very hard rock which on heating gave out a 'nasty smell'. Baron Thénard proposed it; it came from Seyssel, Ain), the Rohart cubes, potassium sulphocarbonate; potassium sulphide pills and carbon bisulphide. They decided that only the bisulphide and sulphocarbonate were of any real use, though neither was perfect and the cost was high, 570 to 645 fr. per hectare.

It was a model report and produced quickly (it is dated 16th November 1877; such speed would be impossible today). It ends with a plea for a renewal of the 2,000 fr. grant to finance the following year's experiments. One sincerely hopes M. de la Rocheterie obtained it.

The phylloxera arrived late in the Champagne and provides another interesting story not without its folly, benevolence and drama. The pest was first officially found at Vincelles (Marne) on 5th August 1890, although the district had considered the matter as early as 1877.[184] M. Gaston Chandon purchased the property and destroyed the vines in an effort to save the district from attack. Although he would receive compensation it would not amount to the loss he thus sustained, except of course that if he had not destroyed the vines at once the phylloxera would have done so, though more slowly. M. Chandon—a renowned Champagne name—deserves praise for his gesture. It was held to be the first time the principle in use in Switzerland (the complete destruction of infected vines and a considerable area round them) was applied in France, though in point of fact similar attempts to exterminate the pest had been made in the Gironde and Burgundy.

In the Champagne a control syndicate was formed; 25,000 people joined it and the body was declared official by a ministerial order of 17th July 1891. It was well organized and sought to control the pest by destroying infected vines, inspecting surrounding vineyards and paying compensation. But in practice it was beset by quarrels and misfortunes and actually achieved very little. According to M. Crespeaux [98b] the syndicate made two serious errors. If it were going to stop the pest it had to act quickly; the syndicate realized this and started pulling out vines before the executive committee had been formed. Many growers resisted the destruction of their beloved vineyards, refusing to believe that their vines were attacked and inevitably would die, thus repeating the old story of other parts of France. As the quarrels developed so did the insects extend and consolidate their grip. To regularize its position and to get some action the syndicate held elections for twenty-five executive committee members on 12th August 1891, the members being able to vote for one or other of two lists—one a members' list and the other an official list, which last had a certain number of members of the Champagne Chamber of Commerce on it.

It would not have mattered very much which list had won, provided action against the pest had continued. As the Champagne is a comparatively isolated vine area the extinction campaign might even have succeeded, at least in delaying invasion for some years. Small growers tended to distrust the large Champagne houses and the belief grew that the object of the Chamber of Commerce was to 'drown the wine-growers'; as a result the *vignerons*' list came out on top in the elections. The local government then committed its second error. The Prefect, possibly annoyed by the election result, allowed six months to pass and only installed the twenty-five syndics on 30th January 1892. He then co-opted twenty-seven new members, mostly from the Chamber of Commerce, who with the residue of eleven and the twenty-five new boys made sixty-three syndics in all instead of thirty-six, much too many. The growers and the grower syndics were furious, the latter resigning *en masse*.

The rump of the committee tried to continue working and did destroy some infected vineyards, but certain *énergumènes* ('tub-thumpers'?) caused trouble, especially when growers were obliged to contribute to the cost of destruction and replanting. Armed resistance to the destruction teams occurred, and there were various brushes with the authorities which, said M. Crespeaux, 'will never be forgotten', so controversial was the subject. Once again, as the humans quarrelled, the insect firmly established itself. Only the usual remedies proved of any use and again the chemicals were too expensive. The area was not without its fantasies. M. Demont (Seine-et-Marne) was a phylloxera/effect man and repeated the view that the French vine had degenerated due to too much vegetative propagation. His cure was to replant with seedlings, a process needing a long time before the plant fruits. Fortunately, in the circumstances, seedling roots are just what the phylloxera likes best and M. Demont's seedling vines were all killed by the pest long before they reached the fruiting stage, thus saving the growers time and expense in proving his theory wrong.

American roots suitable for the chalky soil were needed and eventually were found due to the activities of the Association viticole, formed in 1898, largely composed of those same big firms so attacked by the growers in the first days of the invasion. The vineyards in the Champagne are very closely planted and vast numbers of new plants were needed. Today the area is very prosperous and is firmly planted on American roots, except two tiny spots which are a mystery for most people. Messrs Bollinger have two small vineyards of about half a hectare each, one at Ay and the other at Bouzy, which are on their own roots and produce perfectly normal crops. One can put up a number of theories to account for their survival and knock them all down. They are not deep-rooted in firm compact soil (deep compacted soil might resist penetration by the aphid), but in a light shallow one. They are not in sand, but in a chalky marl. They are not bud sports, which might have thrown a resistant strain, but the ordinary *Pinot noir* used in the rest of the Bollinger vineyards.

M. Ghislain de Montgolfier, of the house of Bollinger, inclines to the view of a soil-effect/genetic explanation. He points out that in the first place the soil is not very suitable for development of the insect and that secondly the system of cultivation practised gives rise to a considerable number of new young roots each season. The growing system used is not to train the shoots to wires but each year to lay the shoot on the ground and cover it with 5 to 10 cm. of earth. In this position the buried second-year wood puts out a considerable number of rootlets and the first (current) year wood, left out of the ground, sends out shoots in the following spring on which leaves and fruit are borne.

Next there is the question of genetic resistance. Boubals [25] found there were two kinds of resistance to the pest; 1, antibiosis and 2, tolerance. In the first case the insect punctures the root with its proboscis but does not like the sap it finds and thus tends to leave it alone; in the second case the insect punctures the plant which then tries to protect itself against attack, in short the classic case of the resistant American vines. M. Montgolfier thinks this last is partly the explanation of the immunity of the Bollinger vineyards. But to many it is still a mystery because it seems to contradict all the now accepted views, such as that marly soils crack easily and have been found to favour the insect, particularly in Germany; because new young roots are just what the creature likes and because up to the present all attempts to communicate root resistance to *vinifera* hybrids have failed, at least to any great degree, so that resistance in a pure *vinifera* is still less likely to occur. Usually the better a hybrid is as regards wine-making qualities the more susceptible it is to the pest (*see* Note 17).

So the Bollinger experience runs contrary to the received views: elsewhere in this book we have pointed out how dangerous it can be to accept these and how doing so in the past set back phylloxera control for some time. We have to keep an open mind on this strange phenomenon and hope the subject continues to be examined. How do these immune patches differ from the rest of the Champagne? Does the soil contain some repellent or insecticidal substance? In the meantime Messrs Bollinger from 1969 onwards have been pressing the grapes from their own-root vines separately and keeping the wine made separate too, so that in 1974 a direct comparison can be made between wines from grafted and own-roots vines from the same area and of the same variety. Will any difference be noted? It will be a test of the greatest interest.

CHAPTER THIRTEEN
Social and Economic Effects
II General

We have now dealt in some detail with four typical areas and turn back to a more general survey on social and economic lines of the attack and cure.

There is a frequently quoted remark of M. A. Lalande's to the effect that the phylloxera cost France more than twice the war indemnity (5,000 million fr.) paid to the Germans, and we need to examine this statement.

Lalande was the president of the Bordeaux Chamber of Commerce and consequently should have been in a good position to assess such figures, though at present we do not know how he arrived at them. He gave a very good opening address to the 1881 Bordeaux International Phylloxera Congress (he was president of its organizing committee), but obviously he could not have given a global figure for losses caused by the pest at this early stage of the invasion. Nevertheless he mentioned costs, inspiring the meeting with the remark that hard work and intelligence would overcome this problem.

Let us briefly examine M. Lalande's loss statement. Dr Guyot [88] estimated that the wine was worth 1,904 million fr. to France annually. The phylloxera killed 40 per cent of the vines over, say, a fifteen-year period. Taking the loss at 20 per cent per year of the above income, because the pest spread slowly and the reconstitution was also slow, we get 381 million fr. per year or 5,715 million for the fifteen years, a figure below M. Lalande's total. But this is not all. Eighty per cent of France's vineyards were eventually reconstituted using American roots. The 1896 total area was 1·8 million ha., so we have to add the cost of replanting 1·44 million of them. At a modest price of 1,500 fr. the hectare (*see* page 164) we get an extra 2,160 million fr. In addition there is the cost of importing wine to make up for some of the deficiency, about 9 million hl. a year at, say, 30 fr. cost and freight, equal to 270 million fr. per year or 4,050 million fr. for the period. We thus have:

Loss of crop, 1878–93	5,715	million fr.		
Replanting	2,166	,,	,,	
Importing wine	4,050	,,	,,	
	12,000	,,	,,	approx.

a figure substantiating that of M. Lalande.

Many other costs should be added to the above estimate, such as the chemicals used in control work, pumping of water, research work, replanting for the second time where the wrong stocks were used in the first place and the cost of currants and sugar used as substitutes for grapes, so that M. Lalande's total is quite moderate. The country in fact lost two wars (the Franco-Prussian and the phylloxera), yet emerged victorious in the end.

Table 9, page 214, shows the wine production in France from 1850 to 1906, with imports and exports and apparent home consumption, from De Grully's figures [53]. We have already mentioned (page 68) the difficulty in assessing the normal yield of vines, but nevertheless this table and Fig. 10 do show both the oidium and phylloxera crises. A word of warning is necessary about the reliability of the statistics. Farmers are notoriously wary of giving away information of this nature, particularly the smaller ones who fear it will be supplying data to the tax man. If a small grower was only making wine for himself and his family (i.e. he was not selling any, at least officially) he would make no return. On the other hand large growers might return more than they had actually produced in order to have an outlet for *piquettes* or other falsified wines. Of course the authorities were aware of this and judged what percentage they must add to the declared total to get what they believed was the true figure. These percentages could be considerable, 18 per cent in 1907, for instance, 11 per cent in 1909 and as high as 28 per cent in 1911.

It seems highly probable that the production figures in column 2, Table 9, include the 'sugar' wines which we discuss below.

The statistics were compiled from the tax returns, and in the early days of the crisis the authorities had no objection to these substitute liquors provided the tax was paid. The natural production of the vineyards themselves is thus the entry of column 2 of Table 9 less the production of sugar wine for 1885–1909 given in Table 10B, page 216. The natural production is entered in the eighth column; in some years the artificial wine production was considerable; 3·6 million hl. in 1888 is an example. By 1909 it had died away (300 hl.).

Wine is an important French crop and always was. Just before the phylloxera crisis Dr Guyot [88] documents it as follows:

Vines are cultivated in	79 Departments
on	2·4 million ha.
producing wine	70·9 million hl.
equal to	29 hl. per ha.
average price	22·97 fr. per hl.
giving an income of	1,628·8 million fr.
providing a living for	1·6 million families, or 6·5 million people one-sixth of the population of France

This income is a quarter of the agricultural income of France produced on a twenty-second part of her surface. These vineyards also produce:

Marc (press cake)	0·2 million tons
giving 50 per cent alcohol (from distillation of marc)	1·2 million hl.
worth	59·6 million fr.
The marc (0·2 million tons) provides cattle food for 80 million rations worth	80 million fr.
and 2·7 million cubic metres of manure worth	16·7 million fr.
The vines produce summer prunings	0·2 million tons
as cattle food worth	28·8 million fr.
The vines also produce winter prunings	0·5 million tons
worth, as firewood	95·4 million fr.

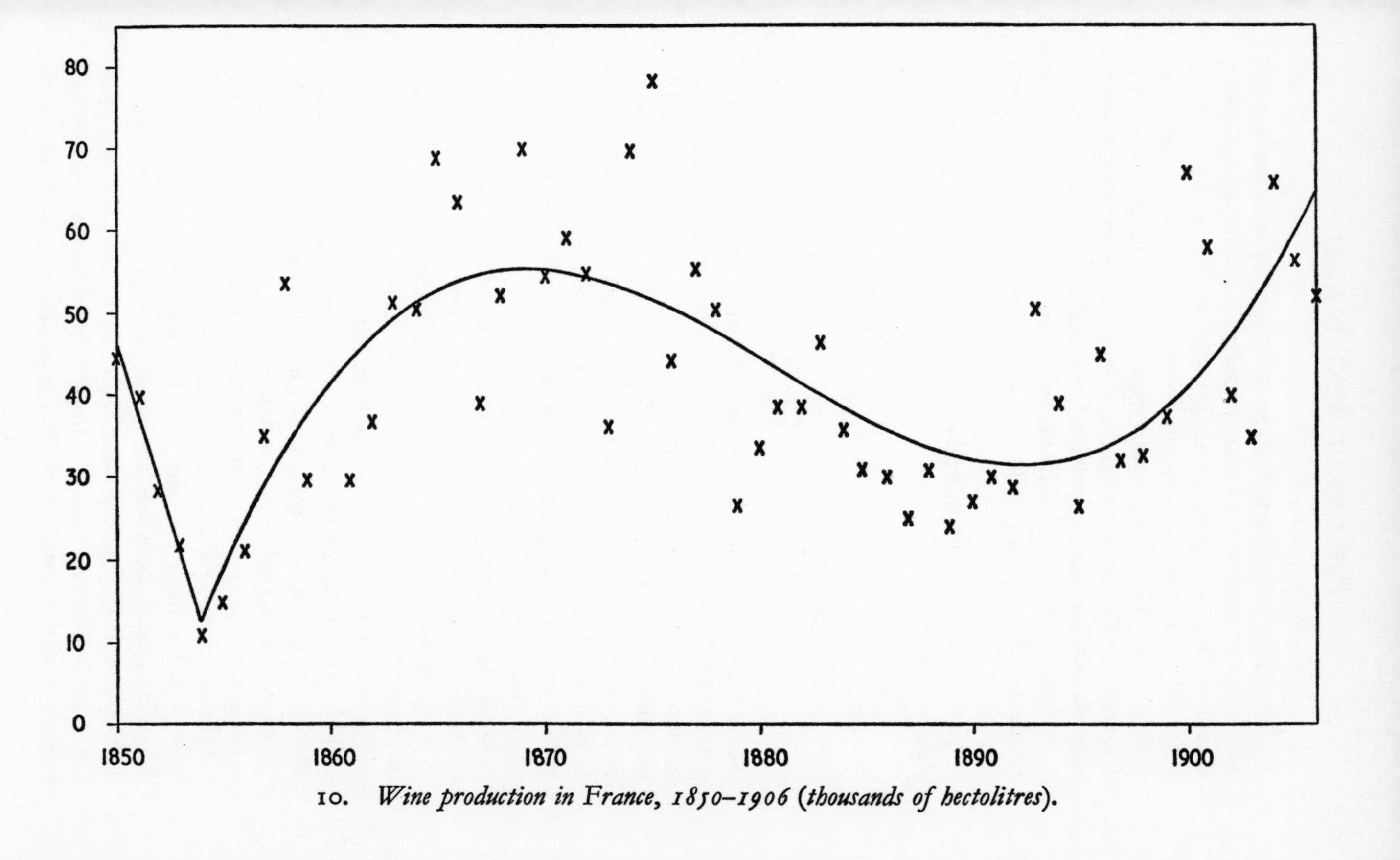

10. *Wine production in France, 1850–1906 (thousands of hectolitres).*

This gives a total production worth 1,909 milllion fr. and provides a livelihood for a rural population of 7·6 million people, a fifth of the population.

In 1896 vineyards were one-fourteenth of the arable area; 1,500,000 growers were spread over 19,000 communes and the wine produced was worth more than a milliard francs, being a seventh part of the agricultural production. In value wine (at 1,174 million fr.) was only exceeded slightly by milk and its products (cheese, butter) (1,182 million fr.) and wheat (1,716 million fr.) [160].

Wine brought a considerable revenue to the state; for the decade 1880–9 this was

Indirect taxes	134 million fr.	
Customs	21 ,, ,,	
Octroi (local taxes)	73 ,, ,,	
	228 ,, ,,	per annum

Taxes from 'other drinks' (mostly brandy, thus derived from grapes) added another 280 million fr. Other indirect taxes were tobacco (370 m. fr.), sugar (101 m. fr.), railways (910 m. fr.) and miscellaneous (114 m. fr.). The wine trade was thus contributing 508 million fr. out of a total of some 2,003 millions, or 25 per cent. Apart from anything else any government would view with alarm some miserable insect threatening to wipe out a quarter of its indirect tax revenue.

In addition to the growers large numbers of people were in the trade. In 1905 M. Ruau estimated them at 1·7 million growers and 1·5 million others in ancillary trades such as bottles, corks, etc.

The two crises show very clearly in Figs. 10 and 11 and Table 9. The oidium caused the crop to drop from 45 million hl. in 1850 to 11 million in 1854 and it gradually recovered, reaching 54 million hl. in 1858 as sulphuring became general. We have no general price figures for this period so cannot really say what happened to total income from wine sales, but one assumes that there was a considerable drop. In the year after the phylloxera discovery (1869) there was an enormous crop and a high price. The crop was 35 per cent more than that of 1868 and the price (29 fr.) only 6 per cent less, for the quality of the wine was good, though Sempé classifies it as 'fair to middling'. Naturally with

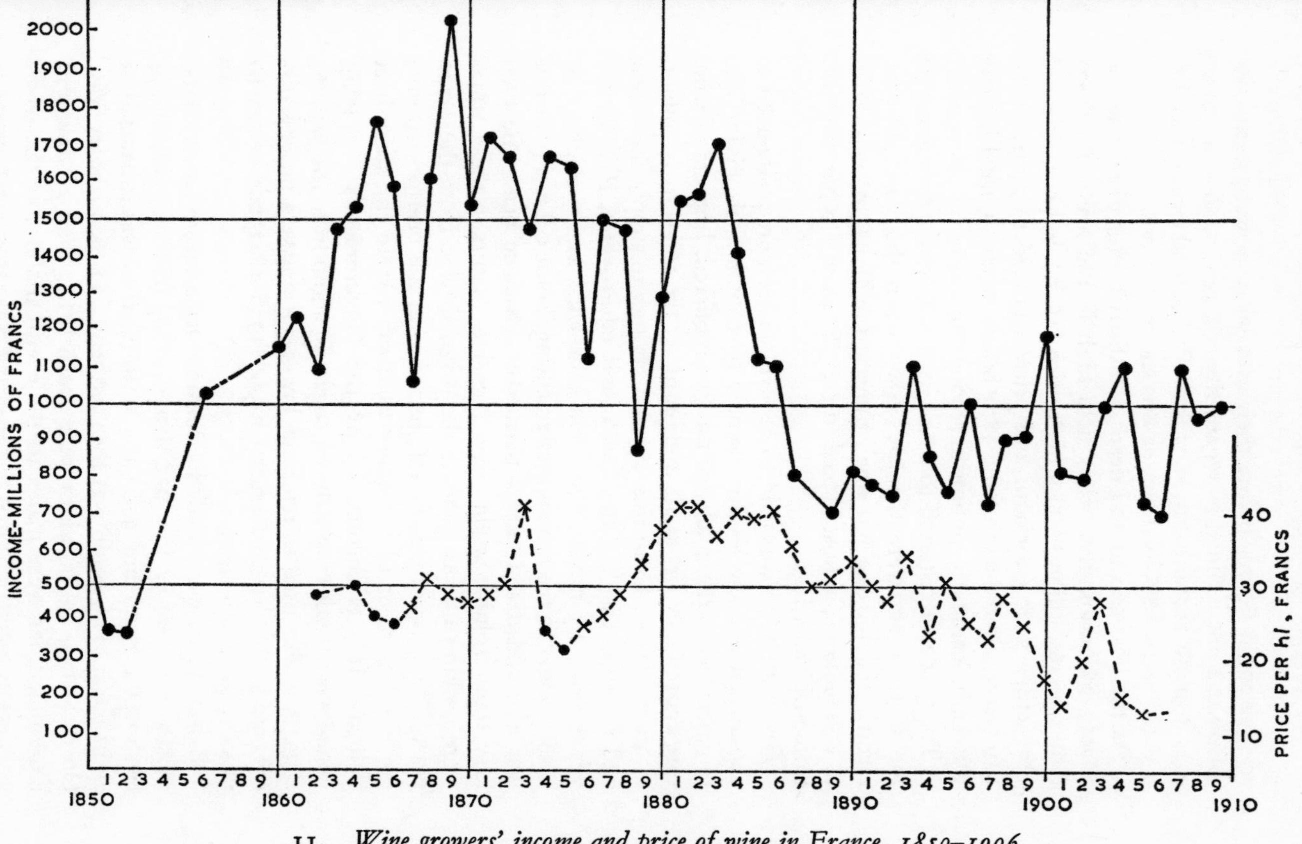

11. *Wine growers' income and price of wine in France, 1850–1906.*

such splendid figures before their eyes wine-growers were optimistic and not inclined to regard the new pest or disease, whatever it might be, with any great fear. The general feeling was that it could easily be wiped out, as was the cattle plague.

As the pest spread and destroyed district after district in the 1870s production and prices held high in the main. The 1873 crop was 36 million hl. at a high price (41 fr.), but the 1874 and 1875 crops were enormous and prices slumped to 21 fr. in the latter year, almost half of the 1873 boom; even so, total income was high, leading to optimism among the unaffected growers.

The pest really started to make itself felt from 1877 onwards; the worst years were the late 1880s when the wine-growers' income of some one and a half thousand millions (fr.) was down to 800 millions, almost halved in fact. Even so, 10 per cent was from artificial wines.

Great attention was paid to exports which were maintained at about their average level of some 2 to 3 million hl., giving rise to anxiety as to the quality of the goods offered. Imports of wine rose enormously, from 0·2 million hl. in the 1860s to 10 million hl. in the 1880s. Even this was not enough to supply the deficit and recourse was had to *piquettes* and raisin wines. *Piquettes* are wines made by mixing the press cake with water, re-pressing, adding sugar to the run and refermenting. Some colouring matter was also added at times, a particular substance being the then new water soluble aniline dye, fuchsine. Naturally these wines were nothing like as good as the original wine from the grape must. The process was called 'bringing in the beetroot to save the vine' as most of the sugar used came from the sugar-beet industry. It was encouraged by the Government's supplying wine-growers with tax-reduced sugar. In addition to the *piquettes* sugar was also used to reinforce low-sugar musts, a process discovered by the French chemist Chaptal (1756–1832) and known to this day as 'chaptalisation'.

Raisin wines were made by soaking dried raisins or currants, mostly from Greece, in water, then drawing off the liquid, adding fresh water, redrawing and so on until the fruit was exhausted of all its sugar. The liquid was then fermented in the ordinary way. On the whole raisin wines were much better than *piquettes* and large quantities were made, to the eventual alarm of the wine-growers. Looked on at first by the growers as a stop-gap to prevent the habit of wine drinking fading away and being replaced by beer, whisky and rum, they were eventually feared as they were found to be so much cheaper than the natural product.

Many people used these processes, and large quantities of these wines were made, as Tables 10 and 11, pages 215–17, show. 260,000 people using 38,000 tonnes of sugar to make 3·6 million hl. of *piquette* (1888) is no mean quantity, and to this must be added the volume of raisin wine made, at least 2 million hl. in that particular year. In the *piquette* table the quality of the first pressing musts is extraordinarily low, around 4 per cent alcohol equivalent, and one is entitled to assume that a certain amount of inaccuracy went into the application forms to enable growers to to secure tax-reduced sugar and thus 'stretch' their vintage; in fact they were using the 'iron cow', so familiar to the fraudulent milkman at that time. But it must be remembered that this was the period of the downy mildew's (*Plasmopara viticola*) first attack, which did so much to reduce quality; nevertheless the low-quality figures persist into the 1890s when control by means of spraying with Bordeaux mixture was well established.

It will be seen that much artificial wine was being made to supplement the lessened natural production.

In 1890 in France we have:

Home production	27·4	million	hl.
Sugar wines	2·9	,,	,,
Raisin wines	3·2	,,	,,
Imports	10·8	,,	,,
	44·3	,,	,,

Thus about 60 per cent of the wine used (including 2 million hl. exported) was home produced, a strange state of affairs for a large wine country. However, France is still a large importer of wine. Home production is now some 61 million hl. a year and about 11 million hl. are imported; thus the home production today suffices for about 85 per cent of her requirements, including exports.

In the late 1880s the large quantities of sugar and raisin wines being made began to alarm buyers, particularly overseas ones. The British Foreign Office was in constant touch with the British Consul at Bordeaux on two grounds, firstly to obtain information about supplies and prices of the good wines Britain was accustomed to import, and secondly to be able to give sound advice to the 'wine-growing colonies and dependencies', such as South Africa and Australia, on phylloxera control. Consul Ward was the gentleman in question and he applied himself to this matter very assiduously, reporting on the pest from its earliest days, as we have already mentioned (see page 67). In 1875 he was hopeful and writes [78]: 'The devastating phylloxera has been very busy. According to the Commission there has been a diminution, that is the phylloxera has only shown itself in two departments hitherto exempt whereas in 1874 four new departments were attacked.' He was making the best of things, as did many others, but that it was still spreading, although not quite so fast, was really but small comfort. In 1878 he mentions the value of wine exports from Bordeaux as £5·4 millions of which Britain takes £1·2 million, prices having dropped from 18*s*. per dozen for cask wines and £1 per dozen for bottles to 11*s*. and 15*s* respectively. By 1883 he is more gloomy: '. . . *low-priced* genuine Bordeaux wines must be regarded as non-existing'. Three years later (1886) he reports on the devastating social effects on the small growers: 'The number, more especially of the smaller class of proprietors on Médoc, in Sauterne and other districts of the Gironde who have been utterly ruined is considerable and the average depreciation of land around Bordeaux compared with ten years ago is estimated at 20 to 60 per cent and even more.' He will have no truck with nonsensical hectolitres or hectares and reports that the 1875 production of 116 million gallons has fallen to 24 million (21 per cent of the former figure); 1875, though, was a year of a big crop. The five-year average 1871–5 is 75 million gallons and thus the 1886 crop was 32 per cent of this average, still a considerable loss.

There was considerable depression among growers and traders and considerable doubt in Consul Ward's mind about the quality of the wine being exported. But he does not despair. In 1885 he writes: 'But without for a moment wishing to deny that this condition of the viticultural interest of France is most deplorable, there would not seem to be sufficient reason for regarding their future prospects with so much very gloomy apprehensions as may be met with at present in this and other parts of the country.' By 1890 he sees his optimism justified, though he seems to have had very little knowledge of the nature of the pest:

> ... more than 150,000 acres ... more or less infected with this noxious worm, and in most other parts of France ... it is nevertheless the opinion of most experienced viticulturalists that the *Phylloxera* may be said to have done its worst ... remedies are being used ... the area of American vines producing direct or of American plants grafted with French vines is annually increasing in a most satisfactory manner.

He found the new mildew affecting quality and was anxious as to whether the copper sulphate spray would affect, even poison, the wine. As for the trade in Bordeaux (about half the shipments from France), it had greatly changed. The quantity shipped had not fallen but old firms could not find quality wines to export. 'A larger share [of the trade] has fallen into the hands of less scrupulous houses, who are less conscientious regarding the genuineness of the liquid which they supply to their customers, many of whom, it is true, often regard not so much the quality of the wine as a handsome and well-sounding label.' In *The Times* of 4th October 1898 [174p] a long letter from Gilbeys, Loudenne, among other things mentions the decline in the English taste for good claret and attributes it to the after-dinner cigarette. The 1860 treaty had brought reasonable wines within the reach of the general consumer though.

The consul had a keen eye for trade possibilities. He records advancing prices for anti-phylloxera chemicals and that the trade in them is increasing. Probably he was confusing the insect and the mildew. By 1893–4 Bordeaux was importing 8,000 tons of copper sulphate for purposes of downy mildew control and most of it came from Britain.

Consul Ward obviously knew the wine trade well. In 1897 he is pointing out how the wine deficiency is being overcome. Wine is being imported from Spain, Italy and Dalmatia, sold 'as is' or blended. Artificial wines also were being made and kept down prices. They filled the gap, but as 'reconstitution' proceeded such practices were starting to alarm growers. By 1897 the American roots were well on their way and Consul Ward reported as follows:

In 1879		Per cent
Seriously attacked	149,000 ha.	82
Unattacked (sandy soils, etc.)	33,300 ha.	18
	182,300 ha.	100

In 1896		Per cent
American or Franco-American replantings	42,300 ha.	31
Constantly treated area	20,200 ha.	15
Abandoned vineyards or grassed down	40,800 ha.	30
Unattacked	33,300 ha.	24
	136,600 ha.	100

So great was the interest in raisin wines and the possibilities of fraudulent conversion that the British Foreign Secretary (the Marquis of Salisbury) in 1890 instructed the ambassador in Paris (the Earl of Lytton) on 7th June 1890 to prepare a report on the subject. Sir J. A. Crowe did this and submitted it to the Secretary on 18th June 1890, thus getting down to the job very smartly. Sir John concluded that 100 k. of raisins or currants would produce 3 hl. of 8 degrees (8 per cent alcohol) wine though some good Greek fruit would give 4·4 hl. Though it is not as good as wine from fresh grapes it is a healthy enough product and a palatable drink. 'In a celebrated case tried in Paris a few years ago, hundreds of samples were tested for the purpose of detecting the difference between wine of raisins and wine of grapes, and four of the best chemists came to the conclusion that, practically, detection was impossible.'[48] One wonders if the court went as far as tasting the stuff!

Sir John gave the following costs for 3 hl.:

Raisins 100 k.	35·00 fr.
Duty	6·00
Freight to Marseilles	2·50
Freight Marseilles to Paris	3·50
	47·00 fr.

or 15·67 fr. the hectolitre. There was, however, a firm in Marseilles offering raisin wines at 15 fr., and according to MM. Jamais and Turrel offers at 12 fr. were made.

The Société Corinthienne, Marseilles, issued a leaflet giving instructions on the making of these wines. In addition to raisins they also supplied a gum for clarifying the wine at 2 fr. the box, sufficient for about 200 l.

There is no doubt that the sugar and raisin wines solved two crises around 1880, for there was a surplus of beet sugar and a shortage of wine, both overcome by the efforts of M. Dubrunfaut in making the one available to the other. The law of 19th July 1880 reduced the tax on sugar from 73·32 fr. to 40 fr. per 100 k. (calculated as refined sugar). Even so, fully taxed sugar was used in 1880 to improve weak musts. The tariff on sugar was put up to 50 fr. in 1884, except for that destined for the *chaptalisation* of wine, cider and perry which paid 20 fr. The following year the law of 22nd July (1885) limited the amounts of tax-reduced sugar each grower might buy; obviously the government was afraid of 'exaggeration'. The quantities were fixed at 20 k. for each 3 hl. of first pressing wine and 50 k. per 3 hl. of second pressings. In order to avoid fraud, that is, the grower reselling the sugar or using it for jam-making or syrups, it had to be tipped into the fermenting vat under the eyes of a tax official, or previously denatured at special depots by being mixed with an equal weight of grapes.

The sugaring allowed the first *cuvée* wines to increase their strength by 4 degrees and the 50 k. per 3 hl. for the second *cuvée* increased the alcoholic strength of these to some 9 to 10 per cent, often making them stronger than the first *cuvée*.

The 20 fr. tax on wine sugar was increased to 24 fr. in May 1887, plus a 'denaturing tax' of 1 fr. per 100 k.

The theory was that sugaring of first *cuvée* wines was simply to supply a deficiency of sugar in a poor year and that the *marc* wines or *piquettes* were just for family use, but sugaring and raisins as substitutes for fresh grapes were bordering on fraud and eventually led to numerous cases of dishonesty in this respect.

The first raisin wine-makers were found in l'Hérault and soon some growers were taking to raisins to make up a deficiency of grapes; there was nothing then illegal in this. To maintain their reputation many growers who used raisins as well would only unload them at night, hoping not to be seen, and the government, worried about this and the revenue it lost from raisin wine not being taxed, passed a law (17th July 1889) making the old cider laws (1816 and 1846) applicable to these wines.

The 1816 law regulated the manufacture of cider and perry in Paris and taxed it and also prohibited the import of cider, apples, pomace or dried apples into wine areas, but so little importance had been attached to raisins that they were not even mentioned;

the making of raisin wine was neither regulated nor taxed. Anyone could do it and gain considerably over taxed, natural wines. Raisin wines thus had three advantages over natural wines: the raw material was cheaper, the freight was less and there was no tax, and as raisin wine was sold as *vin ordinaire*, implying that it was natural wine from fresh grapes, the profits were considerable. By the law of 17th July 1889 raisins intended for wine-making were taxed at the rate equal to the wine tax on 3 hl. per 100 k. of fruit.

Dessert raisins were admitted tax free. Trying to distinguish between dessert and wine raisins gave the authorities much trouble, and many growers found dessert raisins (tax free) cheaper than the bulk wine raisins, spending much time tearing off the fancy packing of the former!

Further legislation (14th August 1889) gave some protection to the consumer, stating that sugar wines and raisin wines must be so labelled, and a law of the following year insisted that any place making these wines must identify itself with a notice board bearing, in 'legible characters', the words 'Raisin Wine Factory'. Obviously there was a rather guilty feeling about this abasement of a noble product. In 1890 *The Times* reported [174m] that there were twenty currant wine factories in Paris alone. The lesser freight and easier packing gave the dried fruit a transport advantage. In fact in times of low prices the wooden barrel itself was worth more than the wine it contained; a barrel-making machine, in the 1880s, brought down prices, however.

Frauds arising from these practices and the complicated legislation trying to control them are many; an interesting one is reported by De Grully [53] quoting Vidal.[183] In 1903 the price of wine from the Aramon vines, at Aimargues, Hérault, averaged 22 fr. the hl: its strength was 7 degrees alcohol by volume. Sugar, with a tax of 25 fr., cost 55 fr. the 100 k. which made 6 or 7 hl. of wine, pointing to the possibility of doubling or trebling one's profit. 'Many', says M. Vidal, 'succumbed to the temptation. The bolder growers did not hesitate, the more timid soon followed their example.' In fact all the authorities cared about was the collection of the 25 fr. per 100 k. of sugar and of 1·50 fr. per hectolitre of wine sold. 'By this time the beetroot was not the enemy of the wine-grower, but his most valued friend, being welcomed with open arms.' At Aimargues between 3rd

September and 26th October the railway station unloaded 53 wagons containing 492 tons of sugar. The official wine crop was 54,450 hl., whereas at least 126,232 hl. (the tax return) were dispatched, an excess of 71,782 hl., for this was the quantity taxed, and no doubt some wine was sold untaxed to friends and relatives as well as being consumed at home.

In Aimargues official applications for low-tax sugar were usually 7 to 8 tons per annum but in spite of the profitability of sugar wines, in September 1903, for example, only ten claims were made totalling 48 tons. With prices at 21 to 27 fr. the hl. growers found it worth while buying sugar at its full price, with the additional advantage that the authorities had no clue as to the amount of sugar wine they made and there was less general neighbourhood publicity as to a man being a 'sugarer'—a maker of poor wine. The 444 tons of sugar used at the full price helped to create a vast quantity of wine—in fact it helped towards the grand total for the country of 15 million hl. of sugar wines, made from 223,000 tons of sugar; this, with the alcohol from the *trois-six* (*see* Note 18) and reserves, led to a market surplus and depressed prices. The Aimargouis people contributed to their own ruin. At first sales had been easy, and many were pleased at being so smart; some were uneasy, but none 'could see that the Tarpian Rock is so near the Capitol Hill'. Prices fell and fell. Soon the victims were clamouring against the early smart-Alecs and soon all were pressing for new laws and their observance. But, says M. De Grully, Aimargues by no means holds the record for fraud. In a nearby commune a grower usually producing 3,000 to 4,000 hl. made 16,000 in 1903. As the statement bears the note '*Cour de Nîmes*, 28 *Juin*, 1905' one presumes he was prosecuted, but we do not know what happened to him. Sugaring, under carefully controlled conditions, is still used in the wine-making countries (*see* Note 19).

Dyes were much used to colour wines both to disguise the dirty white colour of raisin wines or the off-red or dirty-*rosé* of piquettes. One of the favourite dyes was fuchsine (later called magenta), an early aniline discovery. It seemed to arouse the wrath of the press, though the *Journal Officiel* in 1875 [102] pointed out that *pure* fuchsine was quite safe. There was doubt about its purity though, for arsenic was used in its preparation, and the cry was taken up that fraudulent merchants were poisoning the people. The dye was used at the rate of 1 part to 40,000; the safe limit for arsenic is usually taken at 1 part per million, so a fuchsine sample having more than 4 per cent arsenic would give a wine with more than the 'safe' quantity. Four per cent is a very heavy contamination, so in all probability the risks of arsenic poisoning were minimal. *Le Temps*,[172b] in 1876, stated that some of the Midi wine houses were using 10 to 12 tons of fuchsine a year. This would be enough to colour an enormous quantity of wine, say 4 million hl., about the total production of raisin wine and *piquettes*: one feels that the danger was exaggerated. Fuchsine is no longer used in foodstuffs and possibly itself was more dangerous than any likely arsenic contamination.

Elderberries were also used, and it was even worth while to bring in *maqui* berries (the fruit of a Chilean plant *Aristotelia maqui*), the quantities exported to Europe running from 26 tons in 1884 to 431 in 1887,[105] worth $34,541 in this last year; of this last total 316 tons went to France. It was used in Chile for making wine.

The English press, particularly *The Times*, is much worried about sophistication of wines. As far back as 1881 [174h] it had a leading article on the Bordeaux Congress, commenting firstly that the phylloxera appeared to be extinguishing the vine in France as rapidly as Probus introduced it, remarking of the meeting (as true today as then, no doubt) that 'It is the usual fault of such assemblages that the speakers either tell everybody what they already know, or offer opinions and impractical suggestions'. *The Times* then expresses its fears about falsification. In 1882 [174k] there is a report on adulteration: in the previous year out of 3,001 samples examined in Paris 279 were good, 991 passable and 1,731 bad, containing colouring matter such as logwood, fuchsine, beetroot and so on. But later Dubois Frères protested [174l] that though this may happen with some common wines the *grands crûs* are as of high quality as ever and it is costing 300 fr. the hectare to protect them against the pest.

In 1891 we find *The Times* [174n] concerned about cognac supplies. It reports Consul Warburton at Bordeaux as saying that real cognac is scarcely obtainable and that there is much spurious material about. The Charente wine crop used to be 176 million gallons, a figure which had now fallen to 11 million, out of which French consumption had to be taken, reducing still further the amount available for export. A sixteenth of the former production was now being produced, yet shipments had only fallen by half. Obviously the consul was worried by the possibilities of fraud. He had forgotten that it was easier to store cognac than wine, the bulk is so much less, though the evaporation losses are greater. It is likely that some fraud took place, but yet more likely that the cognac houses were unloading stocks and that for them the phylloxera was not a complete disaster. One has but to look at the houses in Cognac itself; the luxurious post-phylloxera buildings contrast greatly with the modest pre-phylloxera ones. It seems that one either perished or made a fortune. Let us return to Consul Warburton as reported by *The Times*. A trial was in progress of a man accused of importing German potato spirit and exporting it as cognac (we do not know how it ended) and, says the consul, M. Jacquemon had produced delicious wines from barley. The secret is to use yeasts from Barsac and Sauterne as they produce wines which distil well. Genuine, old cognac now sells for £2 the bottle, says the consul; an enormous price for those days.

It must be remembered that the press is always avid for scandals about fraud and poisoning of the public, and it is quite likely that wine frauds were much exaggerated. M. Viard (1884) [182] took the view that wine falsification was as nothing compared with the adulteration of other foods, though this was but small comfort if the latter was large. M. Viard quotes what he considers to be a nonsensical article in a serious paper, *Les Débats* of 22nd March 1883, where the author states that wine was doctored as soon as barrelled by the addition of albumen, gum, blood, milk . . . alum, litharge, lead oxide, grain alcohol, fuchsine (contaminated with arsenic), sulphuric acid, tartaric acid, cochineal, orchil and aniline dyes. Obviously the writer knew but little about wine-making as the use of the first four substances is perfectly normal practice. Small amounts of such materials, known as finings, are used to clarify the wine. The other substances might be used at times, but the authorities kept a watch and prosecutions for fraud were not frequent. In the four years 1891 to 1894, in Paris, 202 people were prosecuted (172 by the State, 30 privately). Of these, 55 were acquitted and 187 convicted (why these two figures do not add up to 202 is an unsolved mystery), only three of the convicted being sentenced to imprisonment; the others were fined.

Probably the greatest defraudment of the public was the adding of water at the bar cellar or counter, and of the State in the avoidance of the *octroi* duty. Charles Gide [82] estimated, in 1909, that if each of the wine retailers in France added 5 l. of water a day to their wine stocks they were defrauding the public (and the wine growers) of 8 million hl. a year. He maintains that the only way it could be prevented was to allow the authorities to enter the retailers' cellars. 'But,' he continued 'the Chamber of Deputies would never put its hand on this temple; for them it is sacred.' F. Mommeja has a story of a retailer in the Auvergne being prosecuted for putting 60 l. of water into a barrel (i.e. 25 per cent). Yes, he said, he had watered the wine, but not to that disgraceful extent. He had only put in thirty litres. The judge, much annoyed, said, 'The laboratory knows what it's doing and they say you put in sixty.' The accused's wife then intervened. 'Your honour, my husband is not lying; but you know what men are, so feeble and forgetful. I came behind him and put in thirty litres too.' It was said that heavy freights and taxes forced the bar owner to water the wine.

As regards the *octroi*, the *per capita* consumption figures are of interest. In Paris in 1903 it was 183 l. per head, whereas in the suburbs it was 354 l.[53] Paris would have a number of restaurants and hotels which would tend to increase the *per capita* consumption, not to reduce it. One can only conclude that there was a considerable volume of moonlight flitting past the tax post as well as watering.

Per capita figures can be surprising: wine-growers paid no tax on home consumption and their figure appears to be 666 l. per annum, that is over two bottles a day all the year round for every man, woman and child, a great deal as a steady diet; no doubt much was sold illicitly to neighbours and given away as well.

The point was that fraud had always existed, but the phylloxera crisis threatened the livelihood of so many people that to continue their businesses many more people were tempted to indulge in it and, having fallen, continued more or less in this practice. The equivalent of 'reds under the bed' was seen everywhere. On 25th July 1895 [160] Mr Chancellor, the U.S. Consul at Le Havre, reported to the State Department in Washington, D.C. (September 1895, Consular Reports) that the greater part of the liquid made in France and sold under the name of wine had no connection with the grape, and that in 1895 the Municipal Laboratory in

Paris had destroyed 15,000 barrels of wine because of adulteration. The Bordeaux Chamber of Commerce protested and then the French Government, and the Consular Report was corrected and withdrawn, being qualified as having been caused by 'inexact information'.

The London *Evening Standard*[63] noted that, though the Bordeaux vine area was reduced, prices had not risen, due to wine imports to France and the making of raisin wines. If it got mixed into the clarets exported, said the paper, 'one might as well make the raisin wines on the banks of the Thames as the Garonne'.

The wine shortage in France caused by the pest led to the export of French names, Alicante claret, Californian burgundy, and so on, with which we are so familiar today. A Californian merchant thought the French wines would now be made in California and the Californian ones in France! An allusion to the direct producers in the latter and the grafting in California.

There was a mystique attached to wine-growing felt by the majority of the small growers who found it difficult to envisage any other way of life. Nevertheless there was considerable emigration from the stricken vineyard provinces, the displaced *vignerons* crowding into the towns seeking non-existent work (they had no town skills) and emigrating, particularly to Algeria. At a time when the overall population was increasing (from 1871 to 1891 it was about 6 per cent) the vineyard population declined, as the census figures show. For instance, from 1861 to 1891 the population of Marseilles rose 49 per cent (from 299,256 to 445,784) whilst in the rest of the Bouches-du-Rhône it fell by 11 per cent (from 207,856 to 184,838). In Vaucluse over the same period the population fell by 18 per cent (from 268,255 to 235,411). In the Gironde from 1872 to 1891 Bordeaux increased its population by 26 per cent (381,966 to 480,358), whilst the rest of the province dropped by 3 per cent (323,183 to 313,170). In Algeria the population more than doubled between 1866 and 1901 (from 2·021 millions to 4·739 millions) whereas the population of French origin there nearly trebled (from 122,000 to 364,000).

Bankruptcy figures for the period also indicate the hard times experienced by wine merchants. For instance, a sampling of two months in both 1895 and 1896 shows that respectively 26 per cent and 31 per cent of bankruptcies (of sixty to eighty per month) were wine merchants (wholesale or retail). Another measure of financial stringency is the recourse to pawnbrokers. France possesses considerable statistics in this field—a government monopoly—and they show that loans per annum per 100 inhabitants in towns other than Paris steadily rose from 48·6 in 1875 to 57·1 in 1891, an increase of 17 per cent.

But Gaston Baissette's admirable account [13] of village life in l'Hérault based on contemporary sources epitomizes the misery more vividly than the statistics of emigration, bankruptcy among wine merchants and loans by pawnbrokers. The Fauberge family, living in a village between the plain and limestone hills of l'Hérault, was typical of the small, hard-working proprietors who had built their vineyards over generations. They were wine craftsmen who felt that vines would always exist and produce a crop, and that a man's stability and sense of purpose derived from his attachment to his land; his vines were his lodestone. The family were content with a future of hard, unremitting toil, but they hoped for some steady material improvements for their children, and the Fauberge son had been the first man of his village to travel extensively in other parts of France until he was recalled by his father to fight the phylloxera. The vines had already suffered from oidium (the powdery mildew), cochylis and pyrale (caterpillars) and black rot but none of these troubles had succeeded in wiping them out or threatened them over so long a period as had the phylloxera.

The Fauberges saw their wealthy neighbours buying expensive machinery to flood their land and drown the pest. They themselves tried to find water without success, Auguste, the head of the family, developing an obsessive need to dig a well on his own land. Meanwhile the extensive flooding created mists and caused fevers, rheumatism and malaria.

The vintage failed over a period of years. No longer did the pickers come down from the mountains in the autumn, no longer did the wagons, the horses decorated with rosettes, go to the stations to meet them. The village was like a place of the dead instead of swarming with people, song and fun. Soon grass would be growing in the streets. There were times when there was no meat—no beef, no mutton, no pork. The men of the village mended harness and shod their horses, repaired their houses, stables and cellars and always worked regularly in the vineyards against the time when they would be fruitful again. The Fauberges mortgaged their land to a neighbouring proprietor and the men of the family went to work for him part-time as day labourers, building levees for his flooded land. Even their income from their sheep and silkworms failed as the former died from staggers and the latter from *pébrine*. Years of enormous harvests brought no solace; sugar and raisin wines and fraudulent wines were cheaper to make and good wine therefore commanded few markets. Some of the Fauberges' neighbours left to work the land in the centre and north of France, in New Caledonia, South America and, in particular, Algeria where the Governor-General encouraged the wine-growers to immigrate by intense propaganda. It was not difficult; the wine-growers yearned to regain confidence in the land and swarmed to Algeria with their most precious possession, the vine, and took the pest with them.

Those who stayed behind turned to other jobs while waiting for better times. One man retreated to the hills and set up a distillery, others exploited stone and developed quarries on their properties and made tiles, doorsteps and staircases still in use. One villager made charcoal in the woods; the most humble gathered wood and sold it to bakers, bought chestnut-wood baskets and trugs in the mountains to sell in the villages in the plain. When they got a little money together they planted their vineyards with vines grafted onto American stock. Some of these failed because they did not resist the phylloxera (*see* page 105). At this time (1881) the more resilient villagers emigrated in a second wave; one, at the end of his limits of endurance, hanged himself. The rest sold their remaining possessions of any value.

The Fauberges sold their copper kitchen equipment, their wardrobes and sideboards to the antique dealers and, when the mortgage was foreclosed, they bore with bailiffs with varying degrees of patience. But one bailiff was chased from the village, having seen a painstaking inventory of essential tables and beds torn up before his eyes. He swore he would not set foot anywhere in the district again.

The mood of private bitterness was accentuated by the unimaginative handling of the issues by the Government. Some of the villagers regarded the money prize as an insult typical of government officials who thought that every difficulty could be solved with money and did not deign to discover the problems themselves by visiting the area.

In 1892 Consul Warburton at La Rochelle describes the condition of peasant proprietors as very miserable and says he does not see how the situation can be remedied, for the land taken out of vines will not yield more than 2·5 to 3 per cent and loans cost 5 per cent. Possibly the answer (one we hear advocated today) was the consolidation of farms into larger units. The town was full of peasants seeking work.

There is much testimony as to these miseries from which we will quote some typical examples. Even as early as 1873 Gaston Bazille comments on it.[20] He says it is incredible that people laugh at his fears and at Béziers believe the phylloxera is a 'fabulous beast'. If they would come with him to the Vidourle and other parts of l'Hérault they would be able to see the reality sure enough and a lot of unhappy people, and floods of tears.

In 1878, at Libourne, the Prefect addressed the Viticultural Association pointing out that individual miseries are lost in the general statistics and that he is well aware that many hard cases exist. The commune of Baron, with a population of 500 people, had lost income to the total of 300,000 fr., that is 600 fr. per head, and Puissguin, with 1,000 head, had lost half a million francs. A certain amount of impatience was displayed at the seemingly endless and resultless cogitations of the Phylloxera Commission. 'Send it to the Commission' became a byword for the process of getting rid of a tiresome matter. Even in 1874[172a] *Le Temps* let itself go on the subject, saying that a member of the Phylloxera Commission itself (though not named) had commented somewhat acidly:

> The Commission of 30 has appointed a sub-committee of three members charged with the task of giving the Commission a report on the way in which it would be possible to find a means of permitting it to conceive a procedure by which one would be enabled to propose to the Assembly a method giving events a trend which might lead one to suppose that one had succeeded in producing a combination designed to persuade the public that one had succeeded in drawing up proposals, directed towards a suitable project guaranteed to have the happiest results, without having done anything at all.

The fact was though that the phylloxera was a far more complex matter than oidium control, and the problem was being tackled by both scientists and politicians with reasonable efficiency. For instance, the complex life history of the pest had been confirmed by sending Planchon at considerable expense to America (in 1873); now that the problem of poor growth of American rootstocks in chalky soils had arisen the Government did not hesitate, as we have seen, to send another scientist there. M. P. Viala made a visit in 1888, uncovering a wealth of useful information,[179] the idea having been put forward by the Vigilance Committee of the Charente-Inférieure.

The distress in the Midi led to considerable emigration to Algeria where the new entrants soon set up successful vineyards, exporting their produce to France. The port of Sète was fortunate, for, from being a wine exporting place, it became a wine-importing one, receiving supplies from both Italy and Algeria. The growing importance of France's colony can be seen from the fact that a production of 221,000 hl. in 1873 steadily rose to 8 million hl. in the 1900s. By 1957 it was 15 million, but today independent Algeria is passing out of the French sphere of influence and is tearing up vineyards, possibly inspired by the Mohammedan religion or the realization of a wine surplus market, or both; she is replacing these vineyards with other basic crops.

It was not only the phylloxera which led to distress but also the substitution of cheaper artificial wines to replace the lost production; the artificials then prevented the return of natural wine as the 'reconstituted' vines came into crop.

The new vineyards were better managed, on better land and more heavily manured. Their heavy production together with that of the artificial wines led to a further wine crisis at the turn of the century, the price steadily dropping from around 40 fr. in 1886 to 14 fr. in 1901. Those farmers who by hook or by crook had survived the phylloxera and had replanted on good American roots (frequently after having spent large sums on insecticides, flooding or unsuitable rootstocks), now at the end of their resources, found themselves caught in an over-production period. By political activity they sought to restrain the artificial wines ruining them. A tax strike in the Midi was arranged and an attempt was made to present a petition to the President when he visited the area, though they did not succeed with the latter. Political activity did lead M. Clemenceau to put a duty on foreign wines in 1892, but raisins still came in very cheaply and very little was done to restrain the making of raisin wine until 1897 when the law of 6th April prohibited the sale of sugar wines and *piquettes* and taxed raisin wines at the same rate as other wines. The growers considered themselves saved. Alas! Huge crops from 1898 to 1901 produced as great a crisis as ever and prices continued to fall. However, the next two crops were poor and led to a recovery of prices and, in spite of the law, a vast production of artificial wines encouraged by the price rise and the desperate need to make money to pay off the 'reconstitution' debts. Misery again spread through the Midi for three or four years. Béziers alone lost 20,000 inhabitants, deprived of their living by these frauds. The wine-growers' paper *Le Tocsin* [175] gives a sorry list from which we can consider only a few items—income less than half the normal, often no income at all, not enough food, all credit resources exhausted, children's inheritance sacrificed, trade stagnating, unemployment rife and population leaving. As the vineyards died the Keynesian multiplier effect was seen, for there was less demand for the ancillary products, barrels, corks, clarifying agents, tools, pruners, pickers for the *vendange* and so forth, and these trades too were distressed.

A leader of a strange revolt arose. He was M. Marcelin Albert, and thousands of peasants and even more prosperous men, such as large growers and traders, as well as priests and schoolmasters, joined his movement. It was a growers' revolt with committees, officers, marching songs ('We are the people dying of hunger') and yet with some sound and seemingly complex demands, such as 'Wine must be sold at an average of 1.50 to 2.00 fr. the degree per hectolitre' (that is an 8 degree (per cent) wine should be guaranteed a minimum price of 12 to 16 fr. the hl.). A difficult cry with which to man the barricades! Other, more easily shouted, slogans were 'We believe results when we see them', 'Only natural wine is healthy', 'Results or Revolution'. The elections were boycotted; nobody at all voted and the tax strike continued. Most mayors gave up their posts. A vast assembly of men, women and children began to move on Narbonne; it was said to number 100,000 people and alarmed the authorities. The crowd was orderly, carried the *tricolor* decked with black crêpe, sang 'La vigneronne', its marching song, and camped when and where it could. On 12th June 1907 M. Clemenceau wrote to the mayors asking them to take up office again. The next day the 122nd and the

100th regiments were sent to nearby Larzac. The 'revolutionaries' threw up barricades in Narbonne, but were asked by Ferroul, one of their leaders, to demolish them, and did. More regiments arrived and took the town hall from the rebels. Separatists, Bonapartists and Royalists began to fish in these troubled waters. In Narbonne and many villages troops and wine-growers faced each other and bloodshed was only just avoided. In many cases the 'revolutionaries' welcomed the soldiers with cries of 'Long live the army' and the strains of the *Marseillaise*.

At Narbonne arrests were made, though M. Albert managed to escape. One afternoon the crowd recognized some of the Paris police who had taken people into custody. They seized a policeman called Grosso and carried him to the canal, but were persuaded merely to duck him and let him go. The 136th regiment, alarmed at the course of events, came out of the town hall, bayonets fixed, and fired point blank into the crowd. Many were killed and wounded. The 17th regiment achieved great popularity, for it mutinied, at Agade, and refused to fire on the 'rebels'.

On his own initiative Marcelin Albert went to Paris and obtained an interview with Clemenceau, thinking he could impress the premier with the justice of the wine-growers' cause, although his committee had been against it. How right they were! Clemenceau, that Sunday morning, ran circles round poor Marcelin, showed him that the law must be obeyed, was stern and firm and reduced the wretched wine-grower to tears. Then M. Clemenceau briskly changed his tone; becoming benevolent he said: 'My good fellow, go back home. Calm the people, respect the law, only you can do it. Otherwise you will be arrested.' The threat was still there. 'I haven't even the money for the train' was all M. Albert could reply and Clemenceau gave him 100 fr. The premier was pleased with himself. Twenty minutes to conquer a man who could have raised millions against him. Turning to his secretary he said: 'Let the newspapers know I gave him 100 fr.' Bending over his papers he then ordered the famous 17th regiment to Tunisia and got them out of the way. By the time M. Albert got back to the Midi the press had done its work and he was so reviled by the movement that he gave himself up and went to prison.

The remaining leaders were arrested and imprisoned. The movement achieved nothing directly, but it drew general attention to the problem. The sugar laws of 1904 and 1907 gave the Government powers to suppress frauds and secure the market for the true wines and brandies. Even so the wine economist De Grully [53] was not altogether happy about the future of wine. Writing in 1909, he seems to fear three things as competitors, the bicycle, mineral waters and the motor-car. He says:

> In the same way as the bicycle has killed literature the motor-car is destroying the wine cellar: today the fashionable world seems more inclined to burn paraffin or petrol rather than taste a few good special bottles taken from their hideout. Perhaps the fashion will return, and in any case one should not react too much to the words of certain doctors, not disinterested it is said, whose diatribes against alcoholism too often end up with advice to drink certain waters, always very expensive if not always very pure.

Of course he was writing of the finer wines which, one assumes, he much appreciated. The taste of a wine was its final criterion, after all. Not being one of the 'dismal science' school he recounts an anecdote, attributed to Bertall,[22] on this theme which I repeat here:

> Monsieur X . . ., said to be the most remarkable wine-taster of the neighbourhood, made a point of being able to name the vintage and year of all the classed growths in the Gironde. Any statement he made had the force of law. Never could he be doubted. This art had become a mania with him. He was not a man but just a nose and tongue. He did not live, he tasted. One day his victoria, drawn by a blood horse, hit a heavy cart. M. X . . . was thrown out, fell against a milestone and cut his head open.
>
> He was carried into a nearby house and, whilst awaiting the arrival of the doctor, someone suggested washing the wound with wine. The master of the house brought up a dusty bottle from the cellar, soaked a cloth in the wine and applied it to the cut. A little trickled down the cheek of the dying man into the corner of his mouth.
>
> M. X . . . was still unconscious. Suddenly his nostrils twitched and his lips just moved. Someone bent down to catch his last wishes and heard his dying voice murmur, 'Clos d'Estourel, '48'.
>
> His last words were the crowning glory of his whole life for, of course, he was right.

Another factor tending to help the small growers was the starting of co-operatives. No longer could the middlemen establish their networks of *indicateurs*, small men in the villages who have all the news and local gossip, who can point out to the merchant the growers pressed for money where wine can be bought cheaply, so that a man already in difficulties was further entangled. The co-operatives changed all this by reducing general expenses and being in a stronger position both to sell wine and to buy supplies. They could negotiate serious quantities of wine with shippers and gave sound advice on anti-phylloxera stocks. They were people who counted and no *indicateurs* could affect them.

Another way of looking at the phylloxera problem is its effect on the return on capital. Various authors made studies on the costs of phylloxera control by flooding or insecticides and they run from 250 fr. to, say, 450 fr. per hectare. Edouard Féret in 1878 published an interesting statistical study of wine-growing in the Gironde [69] in which he gives the costs and profits of wine production on a number of sites, mostly before the phylloxera attack. For instance, a *Bourgeois supérieure* vineyard at Margaux of 2·6 ha. cost 3,141·65 fr. to run and returned 3,893·75 fr., a profit of 289·65 fr. per ha. and a return to capital of 5·8 per cent, which, it may be noted, he considers very handsome, saying his readers may well ask why does not everyone put their money into vineyards? In another property studied at Pauillac the profits were 440 fr. per ha. Yet others were:

		francs	
Graves	10 ha.	1,222·35	profit per ha.
Leognan	„	454·10	„ „ „
Côtes Blayais	„	410·80	„ „ „
Entre-deux-Mers	„	190·00	„ „ „
La Brede	„	334·80	„ „ „

The return on capital at the Graves vineyard was 6 per cent, so that the extra expense of 400 fr. for phylloxera control reduced the profit to 772 fr. or 3·8 per cent, but the Graves profit seems exceptionally high. Elsewhere the cost of annual treatment halved or wiped out the profits.

Replanting with American roots was, of course, expensive and we need to examine these costs too. The estimated costs of replanting vineyards vary very much according to the particular ideas of the various writers and their pet theories. Prosper de

Lafitte was a 'winter-egger' of the intensist kind and regarded Americans as expensive folly. The indefatigable Duchesse de Fitz-James was converted to Americanism, but seemed to keep an open mind. She collected a number of figures, including those of a M. Bisset who opposed her, whose account she drily annotates. We may contrast these two gentlemen's figures and refer readers to the duchess's original work for details of her accumulation of more realistic accounts [73] if they desire them (her book is in the library at Kew Gardens). M. Lafitte did not give as much detail as M. Bisset, but nevertheless felt he had established his case against the Americans. 'There are but few *hillsides*,' he says, 'that could support such immense costs.'

	Lafitte fr.	Bisset fr.
Plants	1,250–1,500	480 (*a*)
Planting	500	800 (*b*)
Grafting	150	–
Replacing failures	50	–
Rubbing off roots from scion	–	178
Cost of planting	1,950–2,200	1,458
Interest 4 years on 4,000 fr. (*d*) purchase price [per ha.]		700
Manure		700 (*c*)
Cultivations and interest 4 years		1,276 (*e*)
Pick crop 4th year		44
Pruning		28
		4,206
A *vinifera* vineyard comes to		4,065
Extra for American roots		141

The duchess comments as follows:[73]

(*a*) Expensive, but prices vary greatly.

(*b*) ,, , see my five accounts, varying from 175 fr. to 245 fr.

(*c*) Expensive, 300 fr. is enough.

(*d*) [Bisset put yield at 50 hl. per ha.] The duchess: 'Either the yield was more than 50 hl. or the price was less than 4,000 fr.'

(*e*) Correct. [For details refer to original, page 453.]

However, the duchess was not infallible. She passed an error in the total above (4,196 instead of 4,206), which is not very serious. For 50 hl. the price of wine must be at least 25·90 fr. to meet such costs. But such a vineyard should give 150 hl. per ha. If these figures are true, said the duchess, then the era of wine-growing is over. Obviously she did not believe them which is why she set about collecting the figures mentioned from various other proprietors. Three years' reconstitution, she found, cost from 475 fr. to 685 fr. the ha., which was well within the bounds of financial possibility and the only solution if wine-growing was to continue. And this is what happened; it did continue by means of 'reconstitution'.

Now it would have been quite possible to continue to grow the French vines on their own roots and to give them an annual treatment provided the cost could have been passed on to the consumer, but this was not possible for two reasons. Firstly, various trade treaties would not allow customs' duties to be placed on the import of wine from abroad and, secondly, raisin wines and *piquettes* could be made very cheaply. The French producer was thus powerless in the face of this competition.

Wine began to pour in from Italy and Spain who both saw themselves as the future vineyard of the world and profiting immensely from the French disaster.

Had the wine not come from abroad there would have been a considerable increase in price, so that the consumer might well have turned to beer, cider and cheap alcohol from beet-sugar schnaps of various kinds.

The phylloxera caused floods of ink to flow in the form of technical, political and viticultural articles in newspapers, scientific and other journals; their vast number is an obstacle in writing this book, but the pest has not entered much into literature. It features in Charles Morgan's *The Voyage*[130] where the Maison Hazard vines of Barbet and his mother, Charente Protestants, are continually threatened and finally attacked as Barbet finishes his voyage and sets his prisoners free; perhaps it has a symbolic significance. The pest and the downy mildew are also in the background of François Mauriac's *Destins*,[123] and the work ends with Mme Gornac driving along the road and seeing with great satisfaction that her neighbour's vines are attacked and hers are not.

Robert Louis Stevenson noted the effects of the phylloxera and the despairing attitude to the trouble, and describes a strange drink used as a substitute for wine. In *Travels with a Donkey* [169] he says:

> The phylloxera was in the neighbourhood; and instead of wine we drank at dinner a more economical juice of the grape—*La Parisienne*, they call it. It is made by putting the fruit whole into a cask with water; one by one the berries ferment and burst; what is drunk during the day is supplied at night in water; so, with ever another pitcher from the well, and ever another grape exploding and giving out its strength, one cask of *Parisienne* may last a family till spring. It is, as the reader will anticipate a feeble beverage, but very pleasant to the taste. [At St Germain de Calberte; Gardon de Mialet, St Etienne de Vallée Française, St Jean du Gard.]

In the last chapter 'Farewell Modestine' he says:

> One thing more I note. The phylloxera has ravaged the vineyards in this neighbourhood; and in the early morning, under some chestnuts by the river, I found a party of men working a cider press. I could not at first make out what they were after, and asked one fellow to explain.
>
> 'Making cider,' he said. '*Oui, c'est comme ça. Comme dans le nord!*'
>
> There was a ring of sarcasm in his voice; the country was going to the devil.

One somehow wonders if Stevenson had not got his ideas mixed up and really was hearing about the raisin wines being made in Paris. One cannot really see why the grapes in the '*La Parisienne*' barrel should so obligingly burst one by one and thus extend their fragrance over such a long period.

A man who might have been expected to write on the subject and did not was the great entomologist, Henri Fabre (1823–1915). He lived in the area and, one imagines, would have been intrigued by such a powerful insect and its strange life history. One would like to know why he never mentioned it in the considerable volume of literature he wrote. Almost equally strange is the fact that Tartarin de Tarascon never refers to it. If there was anyone who liked a *bon coup de rouge* it was he, yet, living in the midst of this disastrous threat, he lets it pass unnoticed. Perhaps the suggestion that small birds ate the pests and should not be shot was too great a heresy for this indomitable hunter to comment upon.

I have already referred to Gaston Baissette's remarkable novel, *Ces Grappes de ma Vigne*,[13] which vividly describes these disastrous times in France. Usually people jest about their misfortunes, and the phylloxera, living on the vine, was a gift in this connection. In the Paris *Monde Illustré* of 8th August 1874 a drawing of a somewhat bibulous individual at the agricultural exhibition of that year appeared (*see* Pl. 14). The attendant says to him: 'Monsieur, move away a little; the creature might jump up and seize a nose so full of the good grape!' Between the two world wars, in Vienna, a popular dialect song was 'Die Reblaus' in which the singer infers from his great liking for the Grumholz wine that he must have been a phylloxera in a former existence and that when he dies he hopes to become one again.

Mr Julian Jeffs has kindly drawn my attention to a painting by Salvador Dali, who seems to have been impressed by the phylloxera story, although he can hardly have experienced it directly. Dali comes from Cadaques on the Costa Brava, Spain. At one time the hillsides there were covered with vines; they were all destroyed by the phylloxera and not many were replanted during the 'reconstitution', so that today the once fruitful hills are now just waste land. The theme seems to come into his picture 'The Ghost of Vermeer of Delft which can be used as a table'. On the elongated and horizontal leg of Vermeer there is a bottle and a small glass of red wine. The caption to the picture in *The World of Salvador Dali* [57] reads as follows:

> The alchemists considered dryness and moisture to be the component elements of salt. A spectre that could be used as a table: an eminently eucharistic idea for a painting. A part of the body, at the foot, detaches itself, because of dryness, from the landscape. All the absent wine of the dead vines of Cadaques is condensed in the tiny goblet.

Mr Jeffs adds: 'My understanding from the text is that the fate of the vines had affected Dali profoundly.'

CHAPTER FOURTEEN

Beyond France

The wine trade is highly competitive internationally. When the phylloxera began to devastate the French vineyards several other countries saw it as a splendid opportunity to develop their export trade, at the same time taking measures to prevent the import of the pest. Usually such measures were too late, the insect having entered before the legislation was passed. In this chapter we shall discuss briefly the course of the invasion and its effect in some other wine-growing countries.[29 34]

In most countries the reaction to the discovery of phylloxera was much the same, incredulity tinged with false optimism, followed by panic measures and an inability to learn from the mistakes of others.

CALIFORNIA

There is some confusion about the date of the pest's appearance in California. Davidson and Nougaret [52] maintain that there is evidence for its presence as far back as 1858. Mr H. Appleton [10] definitely found it on the vineyard of a Mr O. W. Craig two miles north of Sonoma Creek on 19th August 1873, though he seems to have written about it only seven years later.

Although the phylloxera was an American insect (with its habitat east of the Rocky Mountains) the Californians were inclined to blame France for the invasion. It seems that Governor Downey, in 1861, appointed a commission to work in the interests of the grape industry and imported twenty thousand cuttings and rooted vines from Europe and Asia Minor and it was thought that the pest had come with these. On the face of it it seems unlikely, for it was not noticed in Europe until 1863, though it may well have been present, and the Californians were bringing in the Catawba from Ohio in the 1850s, especially following Longfellow's publicity for it.

The pest slowly spread (Winkler, 1951) [187] and today is combated by planting in sands or on resistant rootstocks, and some enormous studies have been made as to the best combinations; for instance, the work of Husmann and his colleagues.[95] Over a period of twenty to thirty years they tested 1,903 stock/scion combinations using 6,966 vines and over 300 varieties of grape running from Ach-I-Soum to Zinzillosa, and were thus able to make sound recommendations for all the Californian vineyard soils, doing much to establish the industry on a sound basis.

UNITED KINGDOM

England was the place where the pest was first discovered in Europe. There are no extensive vineyards in Great Britain, and though the insect has been found on living vines from time to time, for instance in 1867, 1868, 1876, 1878, 1883, 1884, 1904, 1907, 1908, 1911, 1912, 1934, 1944, 1956 and 1960, only the 1934 infestation was well established. All invasions were successfully destroyed, the more recent ones by spraying and watering, on the recommendation of the Ministry of Agriculture, with gamma BHC, plus tar-oil winter wash applications in winter. One presumes that the Ministry were not so much convinced 'winter-eggers' as wanting to take no chances. The attacks were mostly in the south-east, the Midlands, Hampshire, Gloucestershire and one in Yorkshire. Two infestations were recorded in Scotland (1876 and 1880) and one in Wales (1878). All infections were in greenhouses except the Hampshire one, which was in an open vineyard. The pest has never been intercepted in the customs.

The planting of open-air vineyards is becoming more popular in Great Britain, and as there is no endemic phylloxera in the country they are nearly all *viniferas* or *vinifera* hybrids on their own roots, consequently very susceptible to phylloxera attack. One urges travellers returning from the continent of Europe not to smuggle vine plants into the country as they could easily bring the phylloxera and do far more damage than a few bottles of, say, illicit brandy! Moreover the plants are not likely to be of much use as English conditions need special varieties. However, Great Britain is not a wine-growing country, and the presence or absence of the pest is of little economic importance.

PORTUGAL (Confirmation 1871, in Bertelo)

The pest did not spread as rapidly as might have been expected, possibly due to a wetter climate, but it did extend and caused alarm in Great Britain about supplies of port wine from the Douro. By 1882, 204,000 ha. were attacked and this had become 300,000 by 1890. It was worst in the north. Out of 100,000 ha., 36,000 were completely destroyed, and in the port wine area 32,000 had gone out of a total of 50,000. By 1895 the whole country was infested. Madeira and the Azores were attacked in due course.

The Government made elaborate arrangements to supply cheap carbon bisulphide to *vignerons*.[150] There was a bonus on the chemical of 25 reis per kilo to viticultural associations and 30 reis to individual wine-growers. Though import duties were waived the sums due had to be deposited and later would be repaid; one assumes, perhaps wrongly, a task fraught with difficulty. The aim was to supply CS_2 at the Marseilles price plus the freight. In spite of this the treatment was too expensive and 'reconstitution' with American stocks was gradually adopted. American species were first used and nearly ruined the wine, then American root-stocks, frequently unsuitable ones, were employed until finally the right combinations were found and all the vineyards (except those in sands) were replanted. Before the pest became general much wine was exported to France (1886, 1·96 million hl.).

TURKEY (Confirmation 1871, at Constantinople)

The pest was first found in the vineyard of the Sultan's Secretary of Grasslands, Kosé Riza Efendi, who had brought in some vines from Bordeaux. By 1886 the whole Constantinople area was infested. Grapes were grown mostly for table and juice purposes.

AUSTRIA-HUNGARY (Confirmation 1872)

The phylloxera was first found in the botanic gardens, Klosterneuberg, and led to a dispute in the papers between Dr Han, an official of the Ministry of Agriculture, and Herr Schoeffel, a member of the Reichsrath. Dr Han brought a libel action against

Herr Schoeffel, who escaped by pleading parliamentary privilege.[34]

By 1889, 219,842 ha. were attacked and 42,000 ha. destroyed by the pest. The Government put up carbon bisulphide factories, selling the material below cost. By the 1890s they were turning to American stock and created numerous nurseries; for instance, in 1894 these establishments delivered 3,086,800 vines and imported 2 million more as well. Obviously they were taking energetic measures, spurred, no doubt, by the fact that up to 1891 they had been a wine exporting country, but had become an importing one from that date, supplies coming mostly from Italy. In the winter of 1894–5, 4·34 million vines were put out (3·86 million from state nurseries), of which 2 million were supplied free of charge. Plantings of ungrafted vines were also made in sands, amounting to 122,722 *jocks*,* a third of the destroyed area. In 1895 the state nurseries supplied 6·63 million vines. The annual production of about 8 million hl. in 1876–80 fell to 4 million in 1892, but the energetic measures taken had restored this to 5·8 million hl. in 1896.

SWITZERLAND (Confirmation 1874, at Prégny, Geneva)

The infestation is said to have started on the estate of Baron Rothschild at Prégny. The baron took active steps against it and destroyed the vineyard, but nevertheless the invasion spread.

The policy adopted was destruction of infection centres with the object of clearing the country completely of the pest. Compensation was paid for destroyed vineyards, and a feature of the campaign was compulsory anti-phylloxera insurance. The destruction was efficient. Two consecutive applications of carbon bisulphide were made at the rate of 150 g. per vine, or say 3 tons per hectare, an enormous rate. Neither was a yet more important feature neglected—examination of the surrounding, supposedly healthy, vines. Vine by vine a square 50 × 50 metres around the attacked point was examined several times a year and if the pest was found the destruction treatment was repeated.

Obviously this was expensive. It was paid for a third by the Confederation (the State), a third by local government (Canton) and a third by a tax on all vine-growers, later the compulsory insurance scheme, and was at the rate of from 5 to 15 fr. per ha.

* We believe a *jock* is 0·5755 ha. According to Martin [122] it is 0·591 ha.

The attacked vineyard temporarily became the property of the State, was surrounded by a fence or cord, had a red flag hoisted on a pole at its centre and a placard with the words '*Vigne sequestrée*' on it (and no doubt in Italian, German or Romance in the appropriate areas). Compensation was paid for two years; in the first year it was the value of the crop, unpicked, plus the value of vines and supports burnt. In the second year compensation was half the value of the last crop.

Another feature of the campaign was the minute details printed and published of the insurance fund and cantonal anti-phylloxera activities. Three examples will show this. The 1892 insurance fund report has a final table comparing 1891 with 1892 (*see* Note 20). There are twelve areas entered; a typical entry is Neuchâtel, 1891. Eight foci were found, consisting of twenty-one attacked vines on 524 square metres. The area destroyed for the whole country only added up to 15,252 square metres, just over 1½ ha. of vines treated in all Switzerland.

The Canton of Geneva's report for 1892 [81] contains particulars of attacks on 160 vineyards with details (all printed) of compensation paid. The highest and lowest of these were 991·25 and 1·60 fr. Sw. respectively and the total compensation paid to 160 people in this canton in 1892 was 16,968·75 fr. Sw., or about 106 fr. per head. It is pleasant to think that the widow Dulac received as much consideration as the 'wealthy' Penays.

Alas! In spite of all this care the phylloxera inevitably spread, much of it, no doubt, coming in on the prevailing wind from France. The country had eventually to turn to American roots, the reconstitution not being completed until about 1951.

ITALY

Although the phylloxera was probably present in Italy in 1870 it does not appear to have been recognized until 1875 or to have become at all general until 1879, when it was found at Lecco and Agrate, Milan Province. The reason for its slow spread was the comparatively isolated nature of Italian vineyards and the habit of growing many vines through trees on long extension shoots. Such plants tend to have deep roots in firmly pressed ground. There is thus no opportunity for the aphids to enter through cracks in the soil and attack the roots.

At first the Italians saw the phylloxera in France as a great opportunity for them. '*L'Italia puo diventare la prima cantina d'Europa.*'[24] Sempé found their statistics difficult. He says: '. . . they are no shining example of fixity and permit all sorts of conclusions to be drawn from them, allowing the viticultural papers on the other side of the Alps great opportunities to make some very strange calculations.' And later he refers to their *bizarreries* and the contradictions in them.

The usual course was followed, it being realized finally that American roots were the answer. At first it left the provision of the new plants to private enterprise and to wine-growers' organizations, but later some government control became essential (Royal Decree of 4th March 1888, No. 2,552 (3rd series), unifying the decrees of 24th May 1874) to overcome the ignorance and fraud prevalent at that time.

Vignerons were in great haste to 'reconstitute' their vineyards and the 'wood-merchants' prospered, 'for the wretched purchaser is not in a position to complain very much until two or three years have passed, if he has been supplied with fraudulent or unsuitable material. Many a bundle of "first-class American wood" in passing from hand to hand changed its variety as many times, now being 420A, now 3309, now 41B, according to local preference'.[4] The State encouraged viticultural associations (*consorzî*) which were easily formed in the north, if they did not already exist, but had to be pushed in the south, and came to exercise more and more control over the sale of rootstocks.

Trained teams were sent out to destroy foci and frequently met with considerable resistance, as in the Côte d'Or, France. The Government bore half the cost of these measures, mostly abandoned during the First World War, which gave the pest a chance to spread. At the end of the war the appalling results of some of the early 'reconstitution' plantings were but too obvious, and energetic steps were taken to regularize the nursery business (Law No. 1363, 26th September 1920). A feature of the earlier campaign was the establishment of a nursery on the island of Monte-Cristo where half a million genuine American plants, true to name and free from phylloxera, were raised and distributed throughout Tuscany. The Government distributed free cuttings of Americans, particularly York-Madeira, and 120 k. of American vine seed and gave subsidies to growers who would establish vineyards with this material.

The Italians produced a number of distinguished phylloxera specialists, such as the famous Professor Battista Grassi, who published an exhaustive study of the genus (thus including other species of *Phylloxera*, such as *quercus*) in 1912.[87] Even at this late date the infestation was not large. Grassi estimated that out of 4·5 million ha. of vines in Italy just under 4 million were still unattacked. But he also points out that this is no reason for complacency. The member of the Chamber of Deputies who maintained that there was no need to worry about or to vote funds for phylloxera defence because France had been attacked, and had overcome the pest by means of American vines, said Grassi, forgot to mention the trifling fact that it cost their neighbour 11 thousand million francs! One did not have to be a prophet, or the son of a prophet, to predict that if steps were not taken the phylloxera would not stop until it had destroyed every *vinifera* in Europe.

In addition to being a great scientist Grassi was a remarkably practical man with an ability to put across his ideas in striking terms. He laid down a successful Italian policy. In 1908 he pointed out that the country did not have the money to destroy the phylloxera; no minister dared ask for the sums needed, which would be at least 100 million lire a year, when the total vote was but 1·5 million, the same sum now with 600,000 ha. attacked as when there were 60. Here he quoted an old saying 'The cake is always the same size and all we get is smaller slices of it'. Even the great German 'success' in Alsace, where they spent a million marks in destruction of foci was all talk. A recent inspection showed Alsace to be infected in spite of the million spent. It was too late to destroy foci one by one because that would not stop the pest spreading. What was needed was for people to know the pest and to delay its attack whilst reconstituting on American roots. The pest was spread by rooted cuttings and plants and never by bare cuttings. The legislation prohibiting the movement of all vines should be repealed and applied only to plants and roots. By allowing the innocuous cuttings to move freely one would avoid the temptation at present existing of smuggling roots around the country and thus spreading the pest. Grassi's hearers were not to think that he was advocating a 'free phylloxera in a free Italy' (a reference to Cavour's slogan, 'a free state in a free Italy') but just common sense. His policy was adopted in essence and, as noted above, considerable control was exercised over nurseries.

SPAIN (Confirmation 1878, at Ampurdán, Gerona)

According to Marcilla Arrazola [118] the insect was first found in Málaga in 1876. It was in Gerona in 1878 and spread to Almería, Barcelona, Granada, León, Orense, Salamanca, Tarragona and Zamora, becoming more or less general by 1888, though Cuenca was not invaded until 1918. Thus, on the whole, the spread of the pest was slow, due probably to the stony and sandy nature of most of the vine soils. By 1889 60,000 ha. had been destroyed. According to Sempé apathy and fatalism gripped growers and municipalities alike—'*Lo que ha de ser no puede faltar*'.* Many of the remedies that had been tried in France and had failed there, or had proved too expensive, were tried again in Spain with similar results. By 1888 the country was turning to American roots, and a Royal Decree (21st August) of that year established a travelling commission, nurseries for American rootstocks and grafting schools at Valencia and Zaragossa. The number of agronomists was also increased, all measures which met with approval, except the final one—a tax equal to 1 fr. per hectare of vines in invaded areas and 0·50 fr. per hectare in other zones.

Much misery spread through the peasant population as the vines died. In the years 1888–9 from Málaga alone 11,000 people emigrated to South America, no doubt assisting the establishment of the Andalucian accent in that area.

Reconstitution of the vineyards was slow, because of the 'terrible interest rate of 8 per cent' demanded by the banks.

The British Foreign Office received reports on the phylloxera from its consuls at various points. Some of these confirm the slow spread and fatalistic attitude. In 1897, from Jerez de la Frontera, comes the following: [78]

> The phylloxera happily appears to be making less progress than was anticipated, but owing to the proverbial apathy which prevails among the majority of landowners in this place, and possibly more so to the lack of capital from which they suffer, little or nothing is being done to prevent the spread of the plague.

* 'What has to be cannot be avoided.'

GERMANY (Confirmation, 1881)

It seems very probable that the phylloxera was attacking in Germany as early as 1875 and we know [34] that between 1879 and 1893 the Empire and States spent the equivalent of 5·7 million fr. on phylloxera control and investigation. At first drastic measures were taken to destroy local infections. These were often bitterly resented by village populations and as late as 1927 there are accounts of resistance to the authorities.[152]

At Hallgarten in that year the pest was found and orders were given to the phylloxera service that the attacked vineyards should be destroyed, compensation being paid at the rate of 70 to 90 pf. per plant which, the growers said, was ridiculous for a seven or eight years' wait before the new, grafted vineyard would crop. The growers were becoming desperate. Excitement ran high in the district. The local control team revolted and asked to be excused from the work of destroying their friends' vineyards. The village thought they had won, but the Phylloxera Commissioner called in a gang of strangers from farther away, some twenty-four to thirty men, and at 8.45 a.m. on Monday, 21st March 1927, they were seen by a youth, approaching from the south. He rushed to the church and rang the tocsin; the people gathered to the cry '*Tzum Reblaus*' and waving their pitchforks they advanced to preserve the vineyards. The team continued to approach, protected by armed police. Fortunately violence was avoided. A parley was held and the work of destruction was stopped whilst the district council discussed a compensation agreement. 'What sadness, what quarrels, tears run down the beards of sturdy men as they see the beautiful vineyards destroyed. And after so much, after war, inflation and taxes, taxes, taxes and now the phylloxera. . . .'

Today American roots are always used when replanting vineyards as the spread of the pest cannot be halted. At Schweigen, near Bad Berzabern, there is a monument to this victory over the phylloxera (*see* Plate 12).

ALGERIA (Confirmation 1885, at Siddi-bel-Abbès)

Up to 1884 Algeria was a net importer of wine, but the heavy immigration from France led to the eventual development of an immense vineyard area. From the first the local authorities were aware of the phylloxera danger, and the law of 12th March 1883 imposed on every grower the duty of reporting any disease symptoms he noticed. This was obviously not enough to ensure vigilance, and every mayor was also required to see that inspectors visited and examined all vineyards in his district every year. In 1884 a law was passed prohibiting the import of vines, fruit and fresh vegetables. It was probably too late by then, or vines and the phylloxera were smuggled in. The pest was found at Siddi-bel-Abbès and at Tlemcen in 1885 and at the other end of the country —Philippeville—in 1886. By 1894 only the Alger department was free of it.

GREECE (Confirmation 1898, at Kapoutzides, Salonika)

The late attack of phylloxera gave Greece a double trading opportunity for wine and currants, reminding one of her classical status as a wine-trading nation. Ancient Greece did reach a remarkable degree of perfection in wine-making. They disinfected their apparatus with smoke and incense, chaptalised it from 1400 B.C. onwards with honey and boiled down juice and pasteurized it from 1000 B.C.[24] Columella[41] said of them: 'Among the ancients there are more examples to follow than errors to avoid' (1st century A.D.).

The Kapoutzides infestation had probably reached Greece from Bulgaria or Serbia. It was frequently expected and some even saw it there before its time. *The Times*, for instance, 19th August 1889, reported its appearance at Morea, from information supplied by its Rome correspondent. This set the diplomatic world in a whirl, the Greek Minister in Rome telegraphing his Government to ask if it was true. All this caused the British Minister in Athens, Sir Edmund Monson, to get a report from Consul Wood at Patras and to write to the Foreign Secretary (the Marquis of Salisbury) on 23rd August 1889 to the effect that the report was untrue and probably had arisen from the fact that a grower 'near Patras had recently requested the government to send an expert to look at his vines which he feared were unhealthy; and that the specialist despatched by the Government had reported that though sickly the vines were perfectly free from the dreaded disease'.[78]

Once again one cannot but reflect how quickly things moved before the airmail swamped all post offices. Four days after *The*

Times statement had appeared Sir Edmund has not only had his man at work in Patras but is also writing a refutation.

Stern efforts, reasonably successful, were made to restrain the phylloxera and B. T. Daris reported[51] that by 1968 out of an area of 220,919 ha. only about 52 per cent was infected. Most of the Peloponnese peninsula, the whole of Crete and Rhodes and many other islands are free of the pest and consequently are much concerned with the possibilities of invasion. Fumigation chambers are provided on the outskirts of free areas, or at posts where all plant material is treated with methyl bromide before being released. At the same time much work is being done to find suitable rootstocks for the different vine areas. It was not possible to maintain this service during the Second World War, and the pest thus spread more rapidly, covering, for instance, the whole of Thessaly.

Control by means of flooding can be done in some areas, such as the Lilantian Plain vineyard in Euboea, but has not proved very effective; it needed vast quantities of water and resulted in low quality wines. Some chemicals have proved effective, the most promising being hexachlorobutadiene; they prove expensive and the biological method of control is obviously eventually the most satisfactory.

The expansion of the currant market for raisin wines ended when France imposed a heavy duty in 1892 and caused a crisis in Corinth.

Reconstitution by means of grafting started in 1922, though nurseries had been controlled from 1914 onwards. As the State nurseries could not fulfil the demand private companies helped too. The Government were enterprising in establishing a nursery and trial ground on the small island of Mykonos (Cyclades) where all stocks were kept free of the insect and true to type. In 1938 another nursery was established at Canna in Crete and in 1948 one in Rhodes as well. The country is thus well placed to provide sound, grafted plants and most vineyards even in pest-free areas are put onto American roots when they come to be replanted.

YUGOSLAVIA (formerly parts of Austria and Serbia and Montenegro)

The area now known as Yugoslavia forms a welcome exception to the usual actions of most countries on being infected by the phylloxera, for the people there took advantage of the knowledge already available in France and the U.S.A. The pest was first found at Pantchevo, near Belgrade, in 1875.[103] It spread to Croatia 1881, Serbia 1882, Dalmatia 1897, Central Dalmatia 1912, Southern Dalmatia 1920 and Macedonia 1912. No attempts were made to use chemicals or wipe out the infestation, the scientists realizing that there was but little chance of doing this with a winged insect, being to the east of a vineyard country (Italy), because the south-west wind would transport it. Flooding was used and much sandy land was planted up as, in fact, it was not much use for anything else, but the main aim was 'reconstitution' on American roots. The first nursery was established at Smedérevo in 1882 and by the end of 1897 there were six in Serbia and twenty-three in Croatia, hence the rebuilding process was very quick. The famous vine school at Klosterneuburg, already mentioned under Austria-Hungary, was of great use, being held by some to have been as influential and helpful as Montpellier.

THE BERNE CONVENTION [44]

This is a convenient point at which to consider the Berne Convention of 1881, an international agreement aimed at preventing the spread of the phylloxera. It arose from a previous convention (17th September 1878) and concerned itself very largely with the question of the transport of plants, with what should or should not cross frontiers, whether plants should or should not carry earth with them and so forth. Obviously the thinking was on the lines of the rinderpest outbreaks, the majority of the delegates either not knowing the phylloxera had wings and could fly, or else ignoring it.

The original countries present were Germany, Austria-Hungary France and Switzerland, represented by a handful of gentlemen with high-sounding names, a general, a baron, a master-of-the-horse, etc. However, the convention was fortunate in having at least one delegate with technical knowledge of the problems, M. Maxime Cornu of France, and later it obtained further help from Portugal and from other agronomists.

In due course it was signed and adhered to as well by Serbia, Luxembourg and Belgium, but, in the very nature of things, could do little to stop the spread of the insect, except to alert the world about its potential dangers. To this extent it may be said to have been successful, for there are still considerable areas of the globe not yet touched by the pest—Crete, much of Russia, Western Australia, parts of California, etc. In addition it drew the world's attention to the need for plant quarantine and led to international conventions for the protection of plants, such as that of July 1949, which have much delayed the spread of pests.

CHAPTER FIFTEEN

Wine and Phylloxera Today

How different is today's wine from the pre-phylloxera vintages? Are the old hands right who maintain that the old wines had an exquisiteness we shall never know again? It is extremely difficult to say because though a few bottles of pre-phylloxera *château* wines can be found from time to time, especially at snob auctions, one cannot compare a bottle a hundred years old with a recent vintage. They are two quite different things.

The problem really resolves itself into two parts: firstly, does the growing of the old varieties on American roots affect the taste, quality and flavour of wine; and, secondly, has the new viticulture induced by the phylloxera crises resulted in different (poorer or better) wine being produced? What is needed to settle the first point is far more blind tastings to see how many people can distinguish a wine from grafted vines from wine from own-roots vines of the same variety and in the same area. The tests need to be carefully made, the subject receiving three glasses, labelled A, B and C. Two glasses contain the same wine and a third the different one. The person then has to pick out the two that are the same and note if there is any difference from the third. The answers, submitted to mathematical analysis, become a statistician's delight. If the two wines tested are of different colours the test must be done in the dark. Such a test will shortly be possible, for Messrs Bollinger have an own-roots vineyard in the Champagne and are pressing the grapes from this area separately. There are also many other areas where grafted and own-roots vines grow very near each other, for instance, Aiguesmortes, France, and many places in Chile and Peru, so similar tests could be arranged with these wines.

The author of this book somewhat doubts if any stock influence in the wine itself will be detectable by the ordinary wine lover.

First, we need to consider what evidence there is of rootstock influence on scion. It is mostly a matter of vigour, of the amount of plant food the root is trying to force into the graft, and there are some striking examples in the East Malling apple rootstocks. These are a series of stocks giving trees from tiny garden dwarfs (No. IX), fruiting in a few years, to great spreading standards (No. XVI), taking ten years to bear a crop. From both trees the fruit itself appears to be indistinguishable, and of course the tree on XVI gives a far bigger crop than the one on IX.

However, one cannot argue that the same is necessarily true of grapes. In 1939 the report [95] of the Husmanns' and Mr Snyder's experiments, already mentioned, included a series of observations on stock effect on grape varieties over a period of thirty years. These workers found slight differences in the dates of the start of vegetation (from 29th to 31st March) and the dates of fruit ripening (from 18th to 24th September) and almost nothing else. A week's earlier ripening could mean a considerable difference in the maturity of the grapes and thus in the resulting wine, the earlier date not necessarily being advantageous as over-ripening could result if the grapes were left too long.

Very little difference has been detected by chemical analysis, but chemical analysis is but a guide to quality, not its criterion.

The new viticulture forced on the world by the phylloxera has resulted in much higher yields per hectare due to planting in richer land and to more scientific management, pruning, cultivation and spraying as well as influence of the stock. The new wines from the new viticulture are thus more likely to be somewhat different from the old and most of them will be superior because the vinification is so much more carefully done today, which balances the extra forcing of the vines by manuring and better cultural care. As to the *grands crûs*, it is an open question and hardly one that can be proved one way or the other because today they are all from grafted vines and can be compared with only a few, very old, surviving bottles, which is not a scientific test, as noted above. The production of vintage wines today is so carefully controlled, both by legislation and by the producers themselves, that the vines are no more forced than in pre-phylloxera times.

We might at this point consider what some of the world's wine experts think on the subject. Some of them have detected differences. In Germany the phylloxera attack has been of much more recent date, in fact it is still continuing, so that more exact comparisons can be made. Only half the German vineyard area at the present moment is on American roots. That great expert on German wines, Mr S. F. Hallgarten, finds [89] that the wines from grafted vines are milder than those from the own-roots vineyards and, as a result, he thinks they will be better liked both inside and outside Germany. Though some of these wines, says Mr Hallgarten, '. . . have entirely lost the characteristics—at least as far as their bouquet goes—of German products, showing that in these cases the American stock was stronger than the German [scion] and decisive for the nature of the wine'.

Mr Hallgarten found that wine from American-rooted vines matured more quickly, often acquiring that 'matured' flavour in one or two years that other wines would get only after five or six. The great André Simon also agreed with this judgment as regards German wines. It is obvious that Germany can effectively fight the phylloxera only by using American roots, and slowly all vineyards are being converted to them. Special methods of vinification and storage are now being adopted to overcome the problem of premature ageing, such as storing in large hermetically sealed containers (glass-fibre tanks are popular), thus doing away with the 'breathing' of the wine through the wood of the barrel, a process bringing about the slow maturing and quality of wine under the old system. Sterile bottling is also popular with the German wine trade. The grafted vines produce bigger crops and live for a shorter time—twenty-five to thirty years compared with double that for the own-roots plants. Much research is being done in Germany on this and other wine problems.

Mr Allan Sichel leaves this matter of the two systems a little more open than does Mr Hallgarten. In the former's fascinating paperback [162] he considers the matter as 'open to discussion', and he continues: 'I have never tasted a wine from an ungrafted vine which has been inferior to the nearest grafted vine, but I have seen examples which are superior. How much this is due to the vine and how much to the soil I do not know.'

Nor is H. Warner Allen [6] very decided on this point. He quotes the early fears when the American roots were first suggested (Bellot des Minières thought there was as much sense in trying to save the French vines by grafting as in trying to milk a he-goat) and comes to the conclusion that the new wines mature earlier and that the grower 'enjoys an increase of quantity, and the wine-lovers have reason to suspect a loss of quality'. He continues, though, by saying that most growers think that, after fifty years of experience, the wines from the grafted vines are showing signs of improvement. Edward Hyams has pointed out that there seems to be a definite amount of quality (bouquet, flavour, taste, depth and so forth, all the terms beloved of wine connoisseurs) to be obtained per hectare and that you can spread this over a considerably varying quantity of wine. You can let the vineyard produce say 15 hl. or force it to produce 150: in the second case you get the quality factors spread over ten times as much wine and must not be surprised to get a wine of much lower value.

We might therefore conclude that today ordinary wines are better than in pre-phylloxera times and that the case of the *grands crûs* wines is unproven one way or the other.

POISONOUS SPITTLE

What exactly does this extraordinary insect do when it pushes its proboscis into a vine root to feed? Why does it kill a *vinifera* and not a *berlandieri*?

In the first place small quantities of saliva are injected into the wound as the insect feeds, and it is the substances that this carries which produce the remarkable reaction. Most of the bio-chemical studies made are more concerned with the differences in the cell content of susceptible and resistant varieties rather than in the nature of the phylloxera saliva. However, in Germany Dr Schaeller [159] analysed the saliva from ten species of aphids and found numerous amino-acids and amides, and Warick and Hillebrant [185] in the U.S.A. identified nineteen of them found in grape tissues. All the gall-forming species Dr Schaeller examined had a substance β indoleacetic acid; none of the non-gall-forming ones had. This chemical is a plant-growth hormone and in fact is frequently used to promote rooting of cuttings. Schaeller was able to induce galls in various plants by applying indoleacetic acid and amino-acid mixtures to them.

As to the different substances in the susceptible and resistant plants, Denisova,[56] in Russia, studied the phenolic compounds there. The root galls on immune varieties contained caffeic acid, quinic acid and glycoside derivatives of quercitin, which inhibit gall formation to some extent. The galls on susceptible varieties contained free quercitin and non-decomposable chlorogenic acid, which both stimulate gall formation. This difference between the vine varieties appears to be innate and not induced by actual phylloxera attack. Denisova also found tannin distribution to be different. In susceptible varieties tannins were evenly distributed, becoming only slightly more abundant in the region of aphid attack, whereas in the resistant varieties the tannins were concentrated around the point of aphid feeding, blocking off the central root cylinder and thus preventing the (poisonous) aphid saliva reaching the sensitive meristem (growing) area. On the susceptible varieties the absence of the concentration of blocking tannins allowed the aphid saliva products to reach the meristem tissues, damage and destroy them and thus the whole plant. Were it not too anthropomorphic we might say that this American aphid, like the American tough guy, could mutter 'Where I spit the grass don't grow'. It is interesting to note in passing that it appears to be the phenolic compounds which inhibit gall formation and that watering with phenols in the early anti-phylloxera tests was said to have had some measure of success (*see* page 99 and Note 10).

Other Russian workers, Dr V. V. Zotov [189] and his colleagues, did some similar work on the differences between resistant (RG) and susceptible (SG) vines and got similar results. They found that the phylloxera saliva contains amylase, proteolytic enzymes and tryptophan. These react on the host starch and protein, depolymerizing them into suitable foods, when growth is stimulated. In the RG varieties induced synthesis predominates and a substance phellogen forms under the damaged tissues, which thus become isolated from the healthy area, and the insects feeding on it starve because the phellogen prevents the passage of food to the aphid zone. In the SG (susceptible) varieties the wounds remain open, secondary bacterial or fungal infections occur and the roots rot. These workers also found that the nucleic acid metabolism was affected by the aphid. Another factor noticed was the respiration rate which, in SG varieties, at first increases with attack and then drops, but in RG varieties continues at an even rate and thus provides energy for growth.

We can thus see that Denisova and Zotov are roughly in agreement. The resistant American plants by selection over the ages have produced a mechanism which isolates the aphid secretion in the first place and is less susceptible to it in the second. Why the insect should want to poison the plant it is difficult to see. Of course the creature does not 'want' to do or not to do any such thing; what is meant is what is the biological advantage to the phylloxera of killing vine roots? Probably none. There undoubtedly is an advantage in forming galls on leaves, as the insects are there protected from enemies. Root galls may also protect the aphids, but death of the host is no advantage to them. Strangely enough, the death of the vine, or poor growth, might be an advantage to the plant species, or to other closely related species, depending on the relative numbers and distribution of plant and insect. If both are relatively scarce over a wide habitat and a vine (say a *labrusca*) becomes infected with phylloxera, from a chance winged migrant, then the plant's death could well lead to the death of all the insects on it and thus to the protection of other *labruscas* in the neighbourhood. Even the *labrusca* becoming sickly could mean the production of far fewer damaging phylloxera and a slower march of the invasion. But if there are many *labruscas* in an area, as in, say, a vineyard planted to Catawba, then one plant infected by the pest and failing under it means that the insects leave the sickened plant and crawl to neighbouring, healthy ones until the whole vineyard is infected and eventually destroyed. It may well be that the aphid secretion producing the protective gall on the leaf, by chance (bad luck) also kills the roots of certain species, to the disadvantage of the aphid. Perhaps some day, say in a million years' time, the aphid, so badgered by man's interference, will have 'adapted' its saliva to the point where it forms the gall and does not kill such species as *vinifera* and *labrusca*. One hopes that man will still be around to see this interesting event if it occurs, but one cannot but feel that the vine and the aphid are more likely to survive this span of time than *Homo sapiens*.

In Australia Miles [126] recently wrote a review article on insect secretions in plants and drew attention to two rival theories on the *cecidogenic stimulus* (the gall-making propensity). The galls, he pointed out, were essentially replicas or simple modifications of normal cells, though in an abnormal relationship to each other. The question consequently is what guides this activity? Bloch [23] wrote that the evidence points towards some sort of chemical influence associated with the salivary secretions which must thus be rather specific for each gall-making insect. Boysen-Jensen[28] thought that the chemical influence was non-specific and that only animals (insects) would be capable of forming complex galls because they alone would be able to control the application of the stimulus, directing it to this or that point according to need and thus shaping the gall to their requirements.

Many other views have been put forward, such as the influence of B vitamins and certain free amino-acids and the whole subject is a field for interesting bio-chemical research. Any reader intrigued by this matter should consult Dr Miles's work [126] as a start. On the whole the evidence available points to the fact that 'IAA' (indoleacetic acid) is the substance leading to gall growth. As the growths always suit the insect, the Boysen-Jensen suggestion, that somehow it directs the application of this growth hormone, seems reasonable.

What is the position of the pest today? To some extent much the same as it was sixty years ago—slowly spreading and being fought with quarantine, insecticides and flooding, while the vines are being reconstituted on American roots.

In most wine-growing countries there are quite appreciable areas where the vines are on their own roots, either flooded every year (*see* Pl. 7) or growing in sandy soil. Some countries and islands have avoided the pest up to the present; Chile, Crete and Rhodes are examples. Others have considerable areas still unattacked, the principal ones being Russia and Australia. Russia takes great interest in chemical control, not so much because American roots are ideologically unsound as because insecticides either appear to be cheaper than replanting (which is very doubtful) or are a stop gap whilst the vineyards are replanted with grafted vines on resistant rootstocks. After many experiments with the usual run of chlorinated hydrocarbon insecticides (DDT, BHC, aldrin, etc.), entomologists in different countries seem to agree that fumigants are the most effective. In Russia, Perov and Mizonova,[139] Kelperis [104] in Greece, and Stevenson [168] in Canada found hexachlorobutadiene to be satisfactory on certain soils. The material had to be injected at 30-cm. intervals, that is about 111,000 injection holes per hectare, a considerable task and one equivalent to the injection of carbon bisulphide in the early days of chemical control. The product employed had 25 per cent of the chemical in it and was used at 230 k. per hectare, thus actually applying 58 k. of active ingredient, a figure showing the product to be much more powerful than the old pesticides, CS_2 and sulphocarbonates. Over double the above-mentioned dose of the new fumigant did not harm the vines in Greece.

Much work has been done on hybridizing vines to produce better varieties, not so much with the object of getting yet better wine as to secure vines with built-in resistance to phylloxera and the two mildews, thus saving growers the expense of grafting and spraying. Considerable success has been obtained with mildew control, resulting in immense economies in buying pesticides and spraying machinery (some of which, such as helicopters, can be very expensive), but so far all Franco-American hybrids have proved susceptible to root attack of phylloxera to some extent, hence they have to be grafted. The hybrids, the 'direct producers', meet considerable resistance from three interests: the established vineyards, particularly the smaller *châteaux* who fear their competition, the government who fear a yet bigger wine surplus (the direct producers tend to be bigger croppers), and the chemical trade who would lose much profitable business. In 1967 France, Italy, Spain and Germany together used about 70,000 tons of sulphur and 80,000 tons of copper sulphate, worth, say £20 millions, nearly all of it used on vines, plus a considerable quantity of other vine pesticides as well.

Strangely enough the planting of hybrids has given the phylloxera another opportunity. The insect does not seem to like *vinifera* leaves and rarely lives on them, though it colonizes the roots and destroys them. On American species it thrives on the leaves and just exists enough on their roots to complete the preliminaries of the sexual stage. The new hybrids show some resistance to the root forms, but the leaves of many of them appear to be attractive to the insect, consequently the pest is again causing economic damage by harming the leaves, not the roots. The pest appears to have fought back and have gained at least one round on points! Spraying against the leaf forms is comparatively easy, using such modern pesticides as BHC, DDT or malathion (Savin,[158] Romania; Stevenson, Canada [167]) but still not as cheap as having varieties immune to all three troubles. Of course it might be possible to find or breed some insect or disease to attack and control the leaf forms of phylloxera but, as we have noted earlier, the insect is particularly well protected. The great Riley found a mite [155] occasionally feeding on the insect in the gall and thought he had discovered a cure, but this was not so.

More recently Nachev [134] found that mites in the soil, far from protecting the plants, actually helped in their destruction as they fed beneath the protecting bark and loosened it.

In order to counter this opposition to hybrid vines growers of them formed a federation known as 'Fenavino' (*Fédération Française de la Viticulture Nouvelle*) with headquarters at Poitiers. In blind tastings wines from hybrid grapes have often carried off prizes. It is not generally known, perhaps, that the popular greenhouse grape in England—Black Hamburg—is a hybrid, the American blood being *riparia*.

It may be noted that the areas growing own-root vines do not make claims, to any great extent, that the wines they produce are superior for this reason, suggesting either that the farmers do not believe they can exclude the pest for very long or that there is no difference between the own-roots and the grafted vines.

It is interesting to speculate what would have happened had grafting *vinifera* onto American species not proved possible. The vine could have continued to be grown in sands and elsewhere with a yearly or twice yearly treatment with insecticides. Sandy soils would have been much sought after and rents or sale value for them would have multiplied. Wine scarcity would have raised prices, at first slowly, but as the pest spread to all the world *vinifera* wine would have become a luxury, and its high price would allow the farmer to pass on the cost of insecticide treatment. However, wine from American species would most likely have been accepted as a substitute for everyday use. One imagines that they would not be immensely popular until at least a generation or so had passed and that the effective vine area of the world would at least have been halved.

Even with the much reduced area of *vinifera* the demand for insecticides would have been enormous and thus the world's chemical industry might have been inspired to do its application research that much earlier and have developed the powerful modern insecticides (BHC, DDT, etc.) only twenty years after their discovery, instead of seventy or more. From this point on one dives into a troubled sea of speculation. The conquest of insect-borne diseases such as malaria or yellow fever much earlier, and also the reaction (resistant strains of pests), and pollution problems too, from the use of these insecticides, are all possibilities, and we might now be in the middle of the population explosion problem, instead of at the start of it. But we would not have had wine as we know it today. It would have been either a great luxury or a rather unpleasant strong-tasting drink. On the other hand it must be noted that some specialists attributed the peculiar and not unpleasing flavour of the 1888 clarets to the widespread use of carbon bisulphide![35]

CHAPTER SIXTEEN

The End of the Story

Phylloxera was *vastatrix*, the destroyer. It would have wiped out the European vine as a common species the world over. In fact it *has* wiped it out as a common species, for *Vitis vinifera* scarcely exists as a wild plant or simple cultivar; it survives only as an artifact, a combination of species, *vinifera* grafted onto resistant roots. As a true species it can only be considered a rare plant found in a few sandy patches and some as yet unattacked places, such as Great Britain, Crete and Chile, an area getting smaller every day as modern transport spreads the pest. Few plant pests completely destroy their host: the only one that comes to mind at the moment being the San José scale (*Aspidiotus perniciosus*, a scientific name almost as alarming as that of the vine pest) and this is usually slower than the phylloxera. Because of its nature phylloxera must destroy or be destroyed, so that for wine to persist it was vital to find a cure or, failing that, to abandon the crop. Men and women responded to this challenge with intelligently applied science. The expedient of grafting, so simple in concept, so hard to achieve, was the final solution which saved wine as we know it for twentieth-century man.

We have demonstrated that phylloxera was unusual in that, introduced to a new continent, it could destroy completely whereas most other insects learned to live with their host and did not wipe it out. Usually pests merely reduce the yield and make the crop more expensive to grow for this reason or by the cost of the measures needed to control them. The phylloxera story shows forcefully that farmers can overcome pests attacking their crops by biological methods of control which usually are as effective as, or more effective than, chemical ones and are much cheaper. It also illustrates the economic consequences which may follow a pest's attack and subsequent control. The protecting of a crop by pest control is always advantageous to the consumer, but under some circumstances may be a drawback to the farmer in that, in time, the measures taken may so increase the crop available to the market as to decrease the price in comparison with that received for the old, attacked crop. Whether this happens or not is a function of the 'demand elasticity' of the particular produce in question, and readers interested in this aspect of the subject are referred to the specialist papers [136] on it.

The cure of the phylloxera trouble appeared to have the anomalous effect of reducing farmers' income. At first wine was imported in large amounts from as yet unaffected areas, then substitutes were used for grapes, and eventually the cure was applied and yield per hectare grew. The substitutes (raisin and sugar wines) persisted so that there was a great overproduction of wine, prices fell by a greater percentage than the crop increase and many farmers were as much ruined by the cure of the pest as they had been by its attack. The general public benefited as they were getting wine of good quality cheaply.

In *Macbeth* the porter, awakened, grumbles

> 'Knock, knock, knock! Who's there, i' the name of Beelzebub?
> Here's a farmer that hanged himself on the expectation of plenty;'

This may just have been comic relief or an observation by Shakespeare that farmers, as a group, might get more money for a short crop than for a big one, the latter being Abel's [1] view (*see* Note 21). The cure, of course, for such situations is not less pest control but some regulation of production. Farms are not quite like factories. The yield is very dependent on an unknown factor—the weather—and this is particularly so with vineyards. Governments now seek to control production to some extent, but it is not quite as easy as it might seem. For instance, one might suppose that surplus wine from good years could be stored to make up the deficiency in bad years. But storing wine in vast amounts means thousands of barrels and tanks, few of which are available. The new crop is always pressing on the accommodation for the old. In the early 1900s' wine crisis the Midi farmers at times merely turned open the taps of the vats to let last year's wine run out to make room for the new, and many wells gave wine and water, intoxicating children and cattle! A favourite device to flatten out violent fluctuations in wine income are schemes to buy up surplus wine and distil it into industrial alcohol, an extremely expensive way of obtaining this product which, for industrial use, can be obtained much more cheaply by the hydration of ethylene from cheap hydrogen sources. Governments naturally dislike buying this surplus and seek to limit the area of vines planted to a figure that will just supply the demand, a matter of some difficulty as production can vary enormously from year to year. Even with such schemes, protest survives to this day. On 4th February 1971 demonstrations by wine-growers at Montpellier, Nîmes and Carcassonne led to battles with the police,[125] reminding one very much of the days of Marcelin Albert. The protest now, as then, was against competition, this time from the wines of Algeria (no longer a colony, but still having the trading advantages of one), and Italy with Common Market rights.

The control of phylloxera and other insect pests needs the continuing vigilance of farmers and scientists; the plenty which results from their control still challenges the ingenuity of economists and politicians to devise a system which will share the advantages equally between producers and consumers.

Our story has shown that the tiny insect found for the first time on a European vine in 1863 in a greenhouse in Hammersmith lived to be a powerful force in Europe. Its conquest was a technical triumph for a devoted band of intelligent men and women of many nationalities who overcame it by the exercise of stubbornness, empiricism, resilience, endless patience, hard work, the occasional flash of genius and blinding glimpse of the obvious. From time to time as we drink our wine we might think of these pioneers busying themselves with experiments in the vineyards in their stiff uncomfortable clothes, their top hats, hard collars, tight waists and long skirts; perhaps we should toast them frequently. They retained the great blessing of wine for us.

St Albans, 1971

Notes

1 (*page 8*). Two other species have since been added to the Muscadinia, all of which have n = 90 chromosomes. The Euvites have n = 20 chromosomes and consequently will not hybridize with the Muscadinia. The Euvites are infertile.

2 (*page 11*). The following is from Pliny's *Natural History* (*c.* A.D. 70) Book XIV. ii, 9.[147]

'Even on account of its size the vine used in early days rightly to be reckoned as belonging to the class of trees. In the city of Piombino is to be seen a statue of Jupiter made of a single vine-stalk that has resisted decay for many ages; and similarly a bowl at Marseilles; the temple of Juno at Metapontum has stood supported by pillars of vine-wood; and even at the present day we ascend to the roof of the temple of Diana at Ephesus by a staircase made from a single vine, grown it is said at Cyprus, inasmuch as vines grow to an exceptional height in that island. And no other timber lasts for longer ages.'

3 (*page 28*). Wars give great opportunities for insect pests to extend their activities. The paramount needs of armies and high commands brush aside all considerations of quarantine regulations. During the First World War the Colorado beetle landed at Bordeaux and started to spread. By the time the conflict was over it was too late to extinguish it and it is now established in most of Europe, except the British Isles. Minor outbreaks occurred in

Great Britain during the Second World War, most likely arising from insects dropped by entomologically and biologically warfare minded airmen. It was not likely to have been a deliberate German policy, as reprisal droppings on German potato fields would have caused greater relative damage, they being so much warmer.

Human lice and fleas flourish during wars. A major mistake of the German High Command was not to use DDT in the armed services as did the Allies. The British medical services, using DDT, were able to say that for every infested allied soldier 'there were several hundreds of lousy Germans'. Dusting the civilian population of Naples with DDT towards the end of the war stopped the typhus epidemic there and saved thousands of lives.

4 (page 30). A number of facts not generally known about C. V. Riley having come my way I thought it advisable to record them here in a note, for, though not directly applicable to the phylloxera story, they do have a general background interest to it and furthermore may help a future biographer of this remarkable man.

Riley married latish in life, in 1878, when he was thirty-five years old. His bride was Miss Emilie Conzelman, the daughter of a prosperous St Louis merchant. They had seven children, Alice, William, Mary, a boy who died in infancy, Helen, Thora and Cathryn. The last three are still alive, two, a doctor and a missionary, both retired and unmarried, living in the eastern United States and the third, a widow, living near London. The youngest girl was given her father's initials C. V., but her middle name was 'Vedalia', the then scientific name of the ladybird which Riley was instrumental in introducing from Australia to overcome a pest threatening to wipe out the California citrus groves. The ladybird has now been renamed *Rodolia* under the priority rules.

After their marriage Charles and Emilie Riley moved to Washington, and in 1889 we find them building a fine home in Wyoming Avenue, Washington Heights. The street was paved and was provided with piped water and sewers and '. . . the residents of Washington Heights do not consider that they are without the modern conveniences merely because they live in the country'. Riley seems to have bicycled considerably, although he kept a carriage. In the summer of 1895 he was returning home with a friend and noticed a new theatre being built. They examined this and in climbing over the new building Riley fell some fifteen feet, cutting his head badly. At this time he was more or less retired, his only official post being that of curator of the museum's insect collection; nevertheless he was always very active. He recovered from his wound and on the morning of 14th September 1895 he was bicycling with his son to Washington when his machine hit a stone; he was thrown to the ground and knocked unconscious, receiving injuries from which he died without in fact recovering consciousness. It was a sad end for an able and much-loved man. He was fifty-one years old. It seems likely that the earlier fall contributed to his demise.

Mrs Riley and her children stayed on in their magnificent house at Washington Heights for two years, the family being helped by Mr Conzelman, her brother. Mrs Riley died in 1946.

Riley appreciated the importance of languages in education, and his children spent some of their youth in France, as he had done, to obtain proficiency in French.

It would seem that Riley had a considerably higher income than one would imagine even a senior civil servant receiving at that time, and his early death does not seem to have left his family in any financial distress.

Riley was very fond of his half-sister, Miss Josephine Lafarge, who became his assistant and secretary. His son William took up the subject of meteorology; he died in 1965 and the eldest child, Alice, in 1968.

Riley in his will, written by his own hand and dated Sept. 29 1888, expresses great affection for both his wife and his half-sister Josephine ('ordinarily called Nina'). The document throws some interesting light on the modesty of his character. He asks for a very modest funeral, though his body may be submitted to dissection if it would advance medical science, the decision to be at the discretion of his wife. He leaves Nina his personal property, exclusive of cash and books, some plots of land in Washington City (Nos. 73 and 74 in Square 239 and the improvements on them, Nos. 1,303 and 1,305, R. Street N.W.) and $10,000. His insects pass to the museum, his books on Hymenoptera to L. O. Howard (his assistant and successor) and those on Coleoptera to E. A. Schwarz. He gives $1,000 to a boyhood friend—H. Wheatley, Walton-on-Thames, England, and $5,000 to his ward Charles F. Wylde, of Southborough, Kent, England, on his coming of age, and he requests Nina to see to his education (Master Wylde was nine years old at the time the will was signed). He appoints Mr Theophilus Conzelman, his brother-in-law, guardian of any of his children who might still be minors should his wife die without appointing a guardian, and leaves all the residue of his estate to her, appointing her sole executrix. One assumes that the sum in question was considerable.

5 (*pages 40, 281*). Although he was born in Bordeaux in 1817 Laliman's family was resident in the island of Santo Domingo. He lost his father and estates in the Negro revolt leading to the establishment of the régime of King Christophe. He and his mother escaped to France and she later married the Marquis of Labrador, the Spanish ambassador at Rome, who remained faithful to King Ferdinand in the Carlist wars. Scarcely eighteen years old, the young Leo Laliman joined the 18th regiment of *cuirassiers* of France and was part of the escort of King Louis-Philippe on the day of the attempted assassination by Fieschi, when Laliman was wounded.

The Marquis of Labrador's close connection with the Carlists led to young Laliman's crossing the Pyrenees to take up a military career in Spain, but the treachery of General Morato ended this and led him to abandon the army. He died in 1897.

6 (*page 49*). One of the minor difficulties the French had with this pest was deciding how to spell the word 'phylloxera', and a few examples of the alternatives follow here. Sometimes it is spelt with an acute accent on the é (*phylloxéra*) and sometimes not, a given author using both variants at different times. Cornu in 1874 has *phylloxéra*,[46] but in 1878 writes *phylloxera*.[47] Planchon also uses both forms, no accent in the scientific journal the *Comptes rendus de l'Académie des Sciences*,[143] and an accent in the literary *Revue des Deux Mondes*.[140] The Duchesse de Fitz-James is also undecided on this point, reversing the Planchon system, no accent in the *Revue* [71] mentioned above and an accent in her booklet on the 1881 Bordeaux Congress.[72] Scientists of that period, such as Balbiani,[15 16] Duclaux,[59] Dumas [61] and Girard,[83] omit the accent, and a scientific journalist, Leroy-Beaulieu,[111] puts it in. The Société nationale contre le phylloxera omits the accent in the title. These are but a few examples; there must be many more. Presumably at some point the Academy made a ruling, for all modern authors and dictionaries now use the accent.

Another curious spelling foible found from that time is Planchon's omission of the 't' in words such as *sarmen(t)s*, *documen(t)s*, *ignoran(t)s*. Obviously it was deliberate and possibly a 'Midi' trademark, though other authors from the same area use the more usual spelling.

7 (pages 64, 121) '*Coulure*' is a condition where much of the fruit fails to set or falls shortly after setting. It is mainly induced by unfavourable weather conditions at the time of pollination or by incompatible or weak pollen. Recent studies assembled by J. B. Free [79] suggest that vine pollination, by wind or insects, is more important than it is commonly thought to be.

8 (page 65). The Government sought first to raise 2 milliards and the 5 per cent loan stock was offered at 82·50, equal to 6·06 per cent yield. More than 4·5 milliards was quickly subscribed.

9 (page 65). The 25-franc gold piece weighed 112·008 grains Troy in 1872. 1 gramme is equal to 15·4 grains, hence the 25-franc gold piece weighed 7·3 grammes; it was 90 per cent fine.

The indemnity paid to the Germans was 5,000,000,000 gold francs.

$$\frac{7{\cdot}3 \times 5{,}000{,}000{,}000 \text{ grammes}}{25} = 1{,}460 \text{ metric tonnes, } 90\% \text{ fine}$$

The area of Alsace and Lorraine together is 1,450,700 ha.

The German dead numbered 44,750; consequently each dead German had gloriously gained his Emperor 32,626 g. of gold and 32 ha. of land, facts which, had he known it, might have consoled him as he died. About £22,000 and 32 ha. of land per dead soldier was an advantageous investment on the face of it, but one does not know what his training and support cost.

10 (pages 71, 99). Professor J. Kennedy, of the Imperial College Field Station, Silwood Park, Ascot, kindly investigated the possibility of systemic aphicidal activity of phenol for me, directing two of his staff, Miss J. A. Ellis and Mr R. F. Skenty, to make some experiments. This lady and gentleman found phenol to have systemic properties in plants but no aphicidal action. They did point out, though, that they were using pure phenol whereas the French experimenters were using commercial carbolic solutions containing many other compounds. It is thus possible that a systemic aphicide may exist in commercial disinfectants but, as noted in the text, it is not very likely or the product would have been more used for phylloxera control in the nineteenth century. I am grateful to the people mentioned above for their help in this matter.

11 (page 75). If the figures for weight of fruit and quantity of sugar from the sulphocarbonate and control plots at Las Sorres are added up (the quantity of sugar being derived from the degrees of the must) we find that the 1205·95 kg. of treated fruit gave 122·44 kg. of sugar and the 689·5 kg. of control fruit gave 64·34 kg. of sugar, the treated almost double in both cases.

12 (page 81). Professor Delpech reported symptoms as described in the text and added a number of others, such as a vast appetite for sexual intercourse and food. We comment only on the latter; 'Patient No. IV: I was so hungry I spent 10 francs on my dinner instead of my usual 6 sous'. Some men became so bad that they could get no other employment and had to stick to vulcanizing and its CS_2 risks.

Prevention by means of better ventilation and treatment with pills containing phosphorus, 'oil', magnesia and (strangely) a few drops of carbon bisulphide were the professor's suggestions. In spite of these difficulties the chemical was employed extensively in the rubber trade, much of it being used to vulcanize toy balloons and 'preventatives' (condoms). It was also used for extracting oils.

13 (page 86). G. Foëx,[75] page 527, here talks of using 50 g. of CS_2 per square metre. We think it must be a misprint for 30, for later he mentions a consumption of 300 kg. per hectare.

14 (pages 100, 108). Hence the success of many 'cures' needing copious watering of some preparation or other. On a failing vine the aphids would have left the plant because of the fall in sap pressure, the water restored the water balance in the plant, which then started to recover, to the joy of both farmer and inventor. As the plant recovered it again became suitable for the phylloxera, was attacked and again failed, probably for good the second time.

15 (page 104). Mr Edward Hyams drew my attention to this little-known fact, documented as follows:

In Dr D. Lucas Alamán's book *Disertaciones sobre la historia de la República Mégicana*, 1844 [5] is an account of the 'Ordenanzas' of the year 1524 dealing mostly with arms (supplies and where they are) and the treatment of the Indians. Page 109 continues: 'Item: que habiendo en la tierra planta de vides de las de España en cantidad que se puede hacer, sean obligadas á engerir las cepas que tuvieron de la planta de la tierra, ó de plantar lo de nuevo, so las dichas penas'.

16 (page 113). At the Bordeaux Congress of 1881 M. Bazille, one of the original identifiers of the phylloxera, also spoke forcefully in support of the Americanists, pointing out that insecticides and flooding, followed by heavy manuring, were but expensive palliatives. Growers should not deceive themselves and try to use apparently cheap remedies, for they would have to come to the expense of replanting with grafted vines in the end. He then recalled a La Fontaine fable which seems to have impressed his hearers. A Midi peasant farmer was fined 100 louis by his seigneur and on the man's complaining that he could not pay this sum his lord said he would propose two other punishments and the peasant could choose one of them. Firstly to eat a dozen garlic bulbs straight off or secondly to receive a beating of thirty strokes. The peasant thought the choice easy: he always ate garlic in any case so he chose the first, but soon his mouth was on fire. 'I must stop!' he cried. 'Beat me instead.' The strongest stableman then started to lambast the peasant's shoulders. After fifteen strokes the peasant cried, 'Stop, stop, I can't stand it. I'll pay the 100 louis,' which he then did. He had suffered all three punishments through not being able to make up his mind to a drastic course of action in the first place. 'See to it,' said M. Bazille, 'that the same thing does not happen in the case of the phylloxera.'

17 (pages 115, 136). As noted in the text, in 1966 M. D. Boubals with the technical help of M. Pistre published [25] some important work on phylloxera resistance, using only the root forms. First of all they established a rapid pot technique for testing vine stocks and varieties and then examined the fundamentals of resistance. The scale of markings was from 0 (very resistant or immune) to 3 (very susceptible) in greenhouse tests and 0 to 5 in field tests. This technique enabled a great deal of time to be saved.

M. Boubals tested all the Vitaceae. Phylloxera were never found on:

Leea guinensis, *Tetrastigma vojneriarum*, *Cissus incisa*, *C. barterii*, *C. discolor*, *C. cactiformis*, *C. quadrangularis*, *C. antartica* (all Class 0).

T. lanceolarium (*C. carnosa*) did support a few insects and a few galls were found on root extremities.

On three species of *Parthenocissus* and twenty-five species and varieties of *Ampelopsis* small numbers of phylloxera could be found at times. (It is thus possible that the pest came to Europe on Virginia Creeper.)

On *Vitis*, as may be imagined, a wide range of susceptibility was found. A somewhat surprising feature was the variability of cultivars within a species. For instance, in *V. riparia*:

17 varieties were classified as 0
41 " " " " 1
1 variety was " " 2

All the *vinifera* were susceptible, but there were degrees of susceptibility, some quite well-known varieties, such as Clairette and Cabernet Sauvignon, showing fewer tumours than others, though all were classified 3. An interesting point is that among the Asiatic vines some were resistant and some were susceptible (*V. reticulata* score 0: *V. davidii* score 3), though none had ever been exposed to the insect in its native habitat. M. Boubals provides a long list giving the susceptibility of hybrids in field trials running from 0 (examples Baco No. 1, Baco 2–16) to very susceptible 5 (examples Excelsior, Ravat 8).

As to the reason for resistance, M. Boubals found that resistant genera repelled the insects and that in resistant species of *Vitis* a corky layer formed beneath the insect punctures, allowing the plant to tolerate the insect to a greater or lesser extent.

M. Boubals issues a *caveat* on accepting American nematode-resistant stocks, as they may not be sufficiently resistant to phylloxera. He found only Salt Creek to be very resistant to the aphid and that Dog Ridge and C.1613 were susceptible.

18 (page 151). Surplus wine was often distilled into spirits of varying strengths, the most popular being *trois-six*. The fraction denotes the quantity of water contained, thus *trois-six*, presumably, had 50 per cent alcohol. There were eleven strengths in all, such as 5/6, 4/5, 5/9, 6/11 etc. but 3/6 was the most popular, its price being quoted in the agricultural press. Makers of liqueurs in the Midi favoured the use of the local *trois-six*.

Why there should be eleven kinds of such spirits is difficult to see. Even in the five mentioned above some differences are very slight and one would imagine scarcely detectable with the rough and ready apparatus then available. For instance 5/9 is 55·5 per cent alcohol and 6/11 is 54·54 per cent. Possibly such close approximations are the reason why only *trois-six* survived.

19 (page 151). Sugaring of wine is still permitted to a limited extent in most countries. In France the 1941 Wine Law prohibits the use of sugar in first *cuvée* wines in certain regions (Aix, Nîmes, Montpellier, Toulouse, Agen, Pau, Bordeaux, Algeria) [39 40] except under certain special circumstances, nor may musts from direct-producer vines planted after 1929 be sugared. In other areas and for second *cuvée* wines the maximum quantity of sugar allowed is 9 k. per 3 hl. of must or 200 k. per hectare of vines, whichever is the smaller. Only dry sucrose may be used, not glucose as this may be contaminated with starch and mineral acids, or syrups which would be a way of 'stretching' the vintage by adding water. Any sugar used attracts an extra tax of 2,000 fr. (1941) per ton. This was the protection the wine-growers wanted, to destroy the profitability of the cheap sugar wines. (The threat of the Greek raisins had been countered by the phylloxera reaching much of that country too.) 1·7 k. of sugar per hectolitre raises the alcohol content of wine 1 degree. Thus the maximum allowable increase of 3 k. per hectolitre means that the alcohol can be raised $\frac{3}{1{\cdot}7}$ $= 1{\cdot}76$ degrees.

In Germany the new Wine Law [90] controls the addition of sugar very strictly. There wine strength is now measured on a weight/volume basis, that is as grammes of alcohol per litre of wine, as distinct from France where the calculation is made volume/volume as cubic centimetres of alcohol per 100 cc. of wine. It must be borne in mind that alcohol is lighter than water, hence the weight/volume figure is less than the volume/volume figure for the same strength of wine. 100 cc. of pure alcohol weigh 79·3 g. Thus French wine of 10 degrees (10 per cent of alcohol by volume, 100 cc. of alcohol per litre of wine) is equal to a German wine of 79·3 degrees (79·3 g. of alcohol per litre of wine).

The new German law says that sugar may be added to must up to a quantity which will not increase the alcohol by more than 30 g. per litre for white wines and 35 g. per litre for reds, and the total alcohol must not exceed 95 g. for whites or 105 g. for reds (equal to 12 and 13·2 degrees on the French scale). Syrups are not allowed, nor are glucoses containing starch. This means that the German wines may have their alcohol increased up to about 3·8 degrees (French) by sugaring.

20 (page 172). Phylloxera Insurance, Switzerland.[11]

Report on area affected 1892, comparing it with 1891. The 'reduction' of the rate of spread by about 0·7 ha. seemed to give great satisfaction.

	1891			1892		
Communes	*No. foci*	*No. vines*	*Area destroyed*	*No. foci*	*No. vines*	*Area destroyed*
Neuchâtel	8	21	524 sq. m.	10	59	634 sq. m.
La Coudre	3	20	100 ,,	5	39	251 ,,
Hauterive	1	7	70 ,,	1	4	30 ,,
Boudry	123	891	4683 ,,	117	1046	3886 ,,
Colombier	49	288	5597 ,,	17	152	1620 ,,
Bôle	14	140	721 ,,	15	48	910 ,,
Cortaillod	3	8	491 ,,	—	—	— ,,
Corcelles	14	145	839 ,,	11	56	374 ,,
Peseux	8	19	412 ,,	4	15	151 ,,
Auvernier	14	126	468 ,,	1	1	10 ,,
Bevaix	12	54	1036 ,,	10	70	373 ,,
St-Blaise	9	33	311 ,,	4	9	133 ,,
Totaux	258	1752	15252 ,,	195	1499	8372 ,,
			1891	258	1752	15252 ,,
			1892	195	1499	8372 ,,
			Reduction	63	253	6880 ,,

21 (*page 192*). The historian of agricultural prices, W. Abel,[1] was well aware of this phenomenon. He says (page 25):

> 'It must not be forgotten, however, that the effect of a good harvest on the farmer's financial profit depended on the level of his marketable production. . . . A passage in Shakespeare that tends to puzzle literary scholars is that in which the porter in *Macbeth* [he then quotes the passage in our text, page 192]. Shakespeare (not the imaginary courtier or scholar but the actor and, as we are told, cornchandler) wrote *Macbeth* in 1603, when grain prices in England had been driven downwards by a series of good harvests. An anonymous author must have had a similar situation in mind in 1767, when he wrote "The farmers are always more afraid of a good year than a bad one. . . . They prefer half a crop to a good harvest."'

Tables

Table 1 Loudon's species of *Vitis*

No.	*Genus*	*Common Name*	*Country of Origin*	*Date of Introduction to Britain*
501	*Vitis* (from the Celtic *gwyd*, a shrub *)			
	Species			
2857	*vinifera* W.	Common		
8	*indica* W.	Indian	Indies	1629
9	*labrusca* W.	Downy leaved Bland's grape	North America	1656
2860	*vulpina* W.	Fox grape	,,	1656
1	*cordifolia* Ph.	Winter grape	,,	1806
2	*riparia* Ph.	Sweet scented	,,	1806
3	*rotundifolia* Ph.	Bull grape	,,	1806
4	*laciniosa* W.	Parsley-leaved	,,	1684
5	*caesia* Sab.	Sierra-Leone	Sierra Leone	1822

* A derivation we much doubt.

Table 2 Genus *Vitis*, The Vines

Section	Group	Species	Origin
First Section:	Muscadiniae	*V. rotundifolia*	(America)
Second Section: Euvites	*a.* Large fruited berries	*V. labrusca*	
		V. candicans	
		V. monticala	
		V. lincecumii	
	b. Small fruited berries	*V. æstivalis*	
		V. riparia	
		V. rupestris	(America)
		V. cordifolia	
		V. berlandieri	
		V. arizonica	
		V. californica	
		V. cinerea	
		V. caribæa	
		V. thunbergi	
		V. flexuosa	
		V. coignetiae	(Far East)
		V. amurensis	
		Spinovitis davidi	
		V. romaneti	
	c. Berries of various sizes	*V. vinifera*	(Europe and Asia)

Table

Showing Method of Presenting Experimental Results,

Plot No.	*Treatment*	*Year*	*Treated Vines*				
			Colour of foliage	Length of shoots (m.)	Weight of grapes (k.)	Degrees of must	Marks, foliage
1	*Rainaud.* Black soap and water applied 6 Mar. 1876. Each vine received 5k. of farmyard manure. Soil compressed on half the plot along N.W. and S.E. diagonal	1876	Very green	1·1 and 1·0	1·75	—	11 10
20	*Bouriol.* Watering with sea water 2 litres ordinary water 5 litres plus a little chalk	1873	Very green	1·2	not weighed	—	12
		1874	do	1·1		—	11
		1875	green	0·9	70·7	9	9
15	*Condat.* Heavy gas oil 1 litre per vine with or without water or 3 to 4 litre alone. 25 Apr.	1873	Fairly green	0·1	not weighed	—	1
		1874	do	0		—	0
		1875	do	0	0	0	0
73	*Louis-Phillippe.* Sealing wax applied hot to pruning cuts	1876	Fairly green	0·5	4·5	8·75	5
57	*Timbal.* Putting at foot of vine chloride of lime 20 g. Pepperwort † 5 g. 27 Feb. 1873	1873	—	—	—	—	5
		1874	—	—	—	—	6
		1875	—	—	—	—	3
		1876	—	0·1	0	—	1

* This column is the marks in column 8 minus the marks in column 13:

† Pepperwort (*Lepidium ruderale*) was at one time thought to be a

— no observations.

3

Las Sorres, Montpellier, 1872–6. Some Selected Entries

Untreated (Control) Vines					*Difference* *	*Remarks*
Colour of foliage	Length of shoots (m.)	Weight of grapes (k.)	Degrees of must	Marks foliage		
—	0	0	—	0	11	Compressed part
					10	Uncompressed part
						Control dead [p. 248]
Very green	1·2	not weighed	—	12	0	2 treated vines and 4 control vines weakened [p. 152]
do	1·1		—	11	0	
green	0·9	57·5	7·5	9	0	
Fairly green	0·4	not weighed	—	4	—4	
do	0·2		—	2	—2	
do	0	0	—	0	0	All treated and control vines are dead [p. 148]
Fairly green	0·5	5·0	9	5	0	One treated vine dead [p. 264]
—	—	—	—	5	0	Two treated and one control vine dead [pp. 124 & 260]
—	—	—	—	6	0	
—	—	—	—	3	0	
—	0·3	—	—	1	0	

if negative the treated vines are worse than the control vines.
remedy for hydrophobia and hence a powerful biological agent.
0 zero result.

Table 4 Experiments at Montpellier 1872–4

Year	*Place*	*No. of treatments made*	*Remarks*
1872 8 May–2 July	Villeneuve-les-Maguelonne	55	Foliage badly attacked by Pyralid caterpillars and tests abandoned
6 July–	Las Sorres (Vigne Sud)	51	4 produced some improvement
1873	Las Sorres (Vigne Sud) (Vigne Nord) (Vigne du Pin)	140	33 produced some improvement 9 were damaging, remainder had no effect
1874	Las Sorres (Vigne Sud) (Vigne Nord) (Vigne de la Chapelle) (Vigne du pin)	 33 28 112 1	 No striking results Sulphides and manure promising Carbon bisulphide caused damage
1875	Las Sorres (Vigne Sud) (Vigne Nord) (Vigne de la Chapelle) (Vigne du Pin)	 100 40 114 59	No treatment destroyed the phylloxera but fertilizers rich in potash and nitrogen with sulphide insecticides, or some salts or soot had some beneficial effect.
1876	Las Sorres (Vigne Sud) (Vigne Nord) (Vigne de la Chapelle) (Vigne du Pin) (Plantier du Puits)	 98 37 110 58 8	Spring frosts badly damaged the vines and made them more susceptible to the pest. Sulphocarbonates and soap showed some hope as insecticides.
		1044	

Table 5 Yield at Mas de Fabre after flooding

Year	*Treatment*	*Area ha.*	*Wine made, hl.*	*Yield per hectare, hl.*
1867	Before attack	15	900	60
68	Attacked	,,	40	2·7
69	2 tons of colza cake, etc., per ha.	,,	35	2·3
70	Flooding (incomplete) (Dec. '69, Feb., Mar. '70) no manure	,,	120	8
71	Flooding (still incomplete)	,,	450	30
72	Flooding: cake 1·75 tons per ha.	20·5	849	41·4
73	Flooding: cake 1·75 tons per ha. (damage by spring frosts)	,,	736	35·9

Table 6 Vineyard Treatments, France, 1880–90

Year	*France Thousands of Hectares of Vineyards Treated*			
	Flooded	*Carbon Bisulphide*	*Pot. Sulpho-Carbonate*	*American Roots*
1880	8	6	1	6
81	8	16	3	9
82	13	17	3	17
83	18	23	3	28
84	23	33	6	53
85	24	41	5	75
86	24	47	4	111
87	n.a.	n.a.	n.a.	166
88	33	68	8	215
89	30	58	9	300
90	33	62	9	436
				1,416

n.a. = not available

Table 7

Resistance to phylloxera (after Millardet)[40]
0 = no resistance. 20 = complete resistance
— = incomplete information

Species	*Resistance*	
	Scale	*Remarks*
Vitis rotundifolia	20	Complete resistance
,, *labrusca*	5	no ,,
,, *vinifera*	0	no ,,
,, *Californica*	–	no ,,
,, *cordifolia*	–	resistant
,, *cinerea*	–	very resistant
,, *candicans*	–	slight resistance
,, *riparia*	18	very resistant
,, *rupestris*	19·5	,, ,,
,, *Berlandieri*	17	resistant
Hybrid 'Noah'	13	fairly resistant

Table 8 History of a typical vineyard, Côte d'Or, 1882–1901. Area of land in *ouvrées* per 50 *ouvrées* total

Year	Area on French roots: Growing	Area on French roots: Dead	Area on French roots: Treated	Area on American roots: Planted	Area on American roots: In production	Total area in production
Fine Wines						
1882	50	—	0·5	—	—	50
3	50	—	1	—	—	50
4	50	—	3	—	—	50
5	50	—	5	—	—	50
6	49	1	7	—	—	49
7	48	2	9	—	—	48
8	46	4	14	—	—	46
9	44	6	22	—	—	44
1890	42	8	35	—	—	42
1	40	10	40	4	0	40
2	36	14	35	8	0	36
3	32	18	26	14	4	36
4	28	22	15	20	8	36
5	24	26	12	26	14	38
6	18	32	10	32	20	38
7	12	38	8	38	26	38
8	6	44	4	44	32	38
9	0	50	0	50	38	38
1900	—	—	—	—	44	44
1	—	—	—	—	50	50
Ordinary Wines						
1882	50	—	—	—	—	50
3	50	—	—	—	—	50
4	50	—	0·75	—	—	50
5	49	1	1	—	—	49
6	48	2	1·75	—	—	48
7	47	3	2	—	—	47
1888	42	8	3·5	—	—	42
9	38	12	5·5	—	—	38
1890	33	17	8·5	—	—	33
1	27	23	10	4	0	27
2	18	32	8·5	8	0	18
3	14	36	4·5	14	4	18
4	10	40	3·5	20	8	18
5	8	42	3	26	14	22
6	6	44	2·5	32	20	26
7	4	46	2	38	26	30
8	2	48	1	44	32	34
9	0	50	—	50	38	38
1900	—	—	—	—	44	44
1	—	—	—	—	50	50

— no data available   source: Laurent [110]

Table 9

Production, value, imports and exports of wine, France, 1850–1906

Year	*Production '000 hl.*	*Price per hl. fr.*	*Value million fr.*	*Imports '000 hl.*	*Exports '000 hl.*	*Home consumption '000 hl.*	*Produc. less sugar wine '000 hl.*
1850	45,266	12·97	587	3	1,910	43,358	—
1851	39,429	10·00	394	3	2,269	37,163	—
1852	28,636	13·14	376	3	2,438	26,201	—
1853	22,062		—	4	1,976	20,090	—
1854	10,824		—	155	1,330	9,649	—
1855	15,175		—	417	1,215	14,377	—
1856	21,294	49·00	1,043	342	1,275	20,361	—
1857	35,410		—	628	1,124	34,914	—
1858	53,919		—	114	1,620	52,413	—
1859	29,891		—	129	2,519	27,501	—
1860	39,558	29·00	1,147	183	2,021	37,720	—
1861	29,738	41·00	1,219	252	1,858	28,132	—
1862	37,110	29·00	1,076	121	1,894	35,337	—
1863	51,362	29·00	1,489	104	2,084	49,382	—
1864	50,653	30·00	1,519	120	2,336	48,457	—
1865	68,913	26·00	1,792	100	2,868	66,175	—
1866	63,838	25·00	1,595	82	3,274	60,646	—
1867	39,128	27·00	1,056	204	2,591	36,741	—
1868	52,098	31·00	1,615	395	2,806	49,687	—
1869	70,000	29·00	2,030	378	3,063	67,315	—
1870	54,535	28·00	1,526	129	2,866	51,796	—
1871	59,024	29·00	1,711	148	3,319	55,856	—
1872	54,920	30·00	1,647	518	3,430	52,008	—
1873	36,000	41·00	1,476	554	3,981	32,573	—
1874	69,937	24·00	1,678	680	3,232	67,386	—
1875	78,202	21·00	1,642	292	3,731	74,763	—
1876	44,306	25·00	1,107	676	3,331	41,651	—
1877	55,273	27·00	1,492	707	3,102	52,878	—
1878	50,634	29·00	1,468	1,603	2,796	49,444	—
1879	26,523	33·18	880	2,938	3,019	26,444	—
1880	33,916	38·18	1,294	7,221	2,488	38,649	—
1881	38,578	40·26	1,553	7,839	2,572	43,845	—
1882	38,825	40·41	1,568	7,537	2,618	43,744	—
1883	46,165	37·01	1,708	8,981	2,541	52,605	—
1884	35,595	40·00	1,423	8,128	2,472	41,251	—
1885	34,481	39·87	1,255	8,184	2,603	37,062	30,694
1886	30,386	40·29	1,224	11,011	2,709	38,688	28,054
1887	25,365	35·98	912	12,282	2,402	35,245	22,478
1888	30,654	30·41	932	12,064	2,118	40,600	22,243
1889	24,032	31·55	758	10,475	2,166	32,341	22,243

Table 9—continued

Year	*Production '000 hl.*	*Price per hl. fr.*	*Value million fr.*	*Imports '000 hl.*	*Exports '000 hl.*	*Home consumption '000 hl.*	*Produc. less sugar wine '000 hl.*
1890	27,416	33·60	921	10,830	2,162	36,084	24,568
1891	30,169	30·06	906	12,278	2,049	40,396	27,079
1892	28,891	28·63	827	9,400	1,845	36,446	26,195
1893	50,703	23·25	1,178	5,895	1,569	55,029	48,036
1894	39,053	23·05	900	4,492	1,721	41,824	37,117
1895	26,688	31·40	838	6,356	1,696	31,348	24,205
1896	44,656	25·27	1,128	8,818	1,783	51,691	40,972
1897	32,351	23·63	764	7,530	1,774	31,107	31,302
1898	32,882	28·88	949	8,625	1,636	39,871	31,130
1899	37,907	25·48	965	8,465	1,717	44,655	36,033
1900	67,353	17·96	1,209	5,217	1,905	70,665	66,512
1901	57,964	14·43	836	3,720	2,022	59,662	57,699
1902	39,884	20·23	806	3,752	1,717	41,919	39,320
1903	35,402	28·12	995	6,189	1,726	39,865	34,547
1904	66,017	16·73	1,104	6,686	1,642	71,061	65,619
1905	56,666	13·23	749	5,171	2,608	59,229	53,389
1906	52,079	13·23	689	5,763	2,110	55,732	51,837

Source: Ministère du Commerce and De Grully.[53]

Table 10 Sugar Wines, France

A. The People

Year	*Total No. of growers '000*	*Number of growers using sugar for:*			
		1st Pressing		*Piquettes*	
		Nos. '000	%	*Nos. '000*	%
1885	1,842	14·4	0·8	33·6	1·8
6	1,785	56·1	3·1	130·4	7·4
7	1,187	59·5	5·0	186·8	15·8
8	1,690	101·9	5·9	160·9	9·5
9	1,688	35·8	2·1	122·6	7·2
1890	1,578	45·4	2·9	170·0	10·8
1	1,578	51·7	3·3	190·0	12·1
2	1,574	36·1	2·3	185·9	11·8
3	1,508	20·7	1·3	128·8	8·5
4	1,552	40·3	2·5	114·6	7·3
5	1,224	34·6	2·8	157·7	12·8
6	1,492	56·7	3·8	143·4	9·6

Table 10—continued

B. The Wines Made

Year	*Sugar used for*		*Equivalent to Wine*		*1st Press. Alcohol increase*	
	1st Press tonnes	*Piq. tonnes*	*1st Press hl. '000*	*Piq. hl. '000*	*from %*	*to %*
1885	2,539	5,394	422	365	3·5	8·7
6	7,095	20,761	973	1,359	4·2	9·0
7	7,656	27,790	1,002	1,885	4·4	9·3
8	12,409	26,354	1,805	1,828	4·0	8·5
9	4,383	15,944	685	1,104	4·7	8·5
1890	6,680	26,388	962	1,886	4·0	8·3
1	8,276	25,673	1,224	1,774	3·9	8·5
2	5,785	22,854	922	1,774	3·6	7·6
3	3,762	14,700	618	1,049	3·5	8·3
4	6,629	13,282	994	942	3·9	8·3
5	6,810	18,585	1,113	1,370	3·6	8·0
6	12,895	18,535	2,344	1,340	3·2	8·1
7	—	—	—	1,049	—	—
8	—	—	—	1,752	—	—
9	—	—	—	1,874	—	—
1900	—	—	—	841	—	—
1	—	—	—	265	—	—
2	—	—	—	564	—	—
3	—	—	—	855	—	—
4	—	—	—	398	—	—
5	—	—	—	277	—	—
6	—	—	—	242	—	—
7	—	—	—	427	—	—
8	—	—	—	321	—	—
1909	—	—	—	188	—	—

Source: *Bulletin de Statistique*, from Sempé [160] up to 1897 and De Grully [53] subsequently. From 1897 onwards statistics were kept of *piquettes* made for home use only: they run from 0·7 million hl. (1901) to 1·8 million hl. (1906).

Table 11

Raisin Wines made in France, 1885–1909

Year	Total hl. '000	Made in factories	
		hl. '000	No. of Manufacturers
1885	2,272	—	—
6	2,820	—	—
7	2,662	—	—
8	2,227	—	—
9	1,826	—	—
1890	3,178	—	124
1	1,704	689	691
2	993	301	142
3	834	299	85
4	544	239	26
5	758	316	26
6	888	—	—
7	451	223	—
8	129	—	—
9	108	—	—
1900	93	—	—
1	38	—	—
2	9	—	—
3	24	—	—
4	52	—	—
5	19	—	—
6	2	—	—
7	3	—	—
8	1	—	—
9	0·3	—	—

Source: *Bulletin de Statistique*, Sempé [160] to 1897, De Grully [53] subsequently.

Appendix

The Prize

From time to time in the course of this work mention has been made of the various prizes offered for a solution to the phylloxera problem. The main one was for 300,000 fr. and was never awarded. It should have gone either to Bazille or Laliman, or have been shared, for they both suggested combining the qualities of the American and European plants by means of grafting, the solution which eventually saved wine for the world.

The first prize offered by the Ministry of Agriculture was in 1870, for 20,000 fr., and the second was approved by the National Assembly, on 22nd July 1874, for the 300,000 fr. mentioned above. Leo Laliman (*see* Note 5) claimed the prize and the correspondence relating to it makes sad reading. The basic reason for refusal was that Laliman had not *cured* the phylloxera, but only found a way of preventing its occurrence, and this was supported by a host of petty reasons as well.

At the session of the General Council of the Gironde, on 1st May 1897, M. Delbay supported the claim of M. Laliman (then just eighty years old) with some heat. He said:

> I know one is often obliged to deny their rights to people who have given great service to their country, and at times one is driven to bargaining with them over the reward they merit. Good Heavens! You could have given gold medals to M. Laliman, in vine congresses you could have proclaimed him the saviour of our vineyards; the newspaper *La Gironde* is a witness of his efforts. And today, when all

he asks of France is that she fulfil the obligations she undertook, to pay the letter of credit now due—the national prize of 300,000 francs —today, when our vines have been saved from disaster by the methods put forward by M. Laliman, today you propose to bargain with an old man who has sunk a large part of his fortune in experiments and hard work which have had the result of saving France's vineyards. You want to swindle this octogenarian of his just reward!

On 4th May 1897 Laliman [84] wrote to the President of the General Council of the Gironde complaining of having been misreported in his claim. He did not say he was the first to introduce American vines to France but that he was the first person to indicate that some of them were resistant to the phylloxera. The president denied that such a mistake had been made.

Laliman had to avoid the implication that he was claiming to be the introducer of American vines, first because obviously he was not—some had been introduced as far back as the sixteenth century—and this could imply general unreliability, and, secondly, he could also be held to be the actual introducer of the dreaded pest, though in all innocence.

Planchon supported Laliman's claim and pointed to Laliman's articles in the journal *Les Vignes Américaines* first mentioning these vines' resistance to the insect.

M. Delbay's words do not seem to have moved the Council of the Gironde to any great extent. Their report of 31st May 1897 says:

> M. Laliman's claim is rejected because all M. Laliman's work, important though it is, had as its object the substitution of American vines or American roots for our French varieties on their own roots. At no time did M. Laliman try to destroy the insect; nor did he try to any greater extent to stop it doing damage: the claims he has made several times to the Ministry or the Agricultural Society are rejected on the grounds that the petitioner has not carried out any of the obligations imposed on him, by the law [i.e. the law establishing the prize]. Consequently I consider the matter need be taken no further.

(The research worker who kindly uncovered this material for me could not restrain herself at this point from adding the gloss 'What a swindle!') [121]

Apparently Leo Laliman died about this time, for the President of the Council added: 'In fact the *dossier* can be withdrawn because the petitioner no longer exists.' The Prefect agreed, but added '. . . but with the reservations noted.'

The reservations apparently gave Laliman's son an opportunity of re-opening the claim, which he did in a letter dated 19th April 1898:

> . . . I request kindness and justice from the General Council of the Gironde and the distinguished help it can give me in order to obtain national reward earned by my father, the indefatigable pioneer who, from the moment of the invasion of our vineyards by the phylloxera, and even at the cost of the greatest sacrifices, has so greatly contributed to their reconstitution and health. . . . The General Council itself, for twenty years, has recognized that the only hope of survival for our vineyards menaced by the phylloxera is the use of American vines. . . .

Laliman *fils* was no more successful than his father. Further research might uncover some reason for the refusal of the prize, but in the meantime one can only assume that it sprang mainly from the reluctance of governments to part with money, helped by the general idea in official circles that a chemical control was really needed, because there would then be considerable insecticide sales, equal to or in excess of the sulphur and copper sulphate sales for control of the two mildews, and the haunting suspicion that it was Laliman who brought the pest in the first place. Furthermore Laliman, with his Carlist background, may have been politically suspect by the republican government, but this is mere speculation. The prizes were never awarded and thus saved the government 320,000 fr.; not a large sum seeing that they were dealing in billions.

Prizes of this nature seldom achieve their object. Some twenty years ago a reward was given in the Caribbean for a banana plant resistant to the devastating Panama disease, with the opposite result of the Laliman case. The money was paid, but the banana was not particularly resistant.

The French offer was beneficial in the end. It led to such a flood of ideas that the Montpellier research work had to be extended to test them, leading to the final solution and a far better understanding of the whole subject. Montpellier is justly proud of this work and has erected two statues to it; one to Planchon and the other to Foëx, the latter symbolizing the rescue of the ailing French vine by a nubile young *Américaine* (*see* Pls. 10 and 11).

Bibliography and References

1. ABEL, W. *Agrarkrisen und Agrarjunktur*. Parey, Hamburg-Berlin, 1966. (English translation, by Olive Ordish, in the press. David and Charles, Newton Abbot.)
2. ADLUM, J. *A memoir on the cultivation of the vine in America*. Washington, 1823.
3. AGUILLON, —. *Rapport presenté par M. G. Vionait, vice-président du comice agricole d'Epernay*. Commission internationale de viticulture, 1869.
4. AKENHEAD, D. Memorandum No. 11, Empire Marketing Board. H.M.S.O., 1928; p. 20.
5. ALAMÁN, D. LUCAS. *Disertaciones sobre la historia de la República Mégicana*. Mégico (Imprenta Mariano Lara), 1844. Tomo I, Ordenanzas del año 1524. Originally published by Srs Fernandez, Sabrá and Sainz de Baranda. Academía de la Historia, Madrid, 1843.
6. ALLEN, H. WARNER. *Natural red wines*. Constable, London, 1951; p. 57.
7. *L'Année scientifique et industrielle*. 18e année, Paris, 1874; p. 366.
8. *Annuaire de la noblesse de France* (ed. Borel d'Hauterive). Paris, 1885; p. 67.
9. APPEL, O. Zur Kenntnis der Überwinterung des Oidium Tuckerii. *Zentralbibliothek für Bakteriologie und Parasitologie*, Abteilung II, Vol. II, p. 143. Jena, 1903.
10. APPLETON, H. *The* Phylloxera vastatrix *and its ravages in*

Sonoma Valley. Appendix C. Board of State Viticultural Commissioners, California, 1880; pp. 108–11.

11. *Assurance mutuelle contre le phylloxera*. Rapport de la Commission Administrative, 1892. Neuchatel, 1893.

12. BADELL R., D.L. Vides Européas. *Anales de la escuela de peritos agrícolas . . .*, XI, pp. 91–232. Barcelona, 1952.
13. BAISSETTE, G. *Ces grappes de ma vigne*. Editeur Français Réunis, Paris, 1956.
14. BALACHOWSKY, A., and MESNIL, L. *Les insectes nuisibles aux plantes cultivées*. Paris, 1935; p. 713.
15. BALBIANI, E.-G. Mémoire sur la reproduction du phylloxera de la chêne. *Mémoires présentés par divers savants à l'Académie des Sciences de l'Institut National de France*, XXII, 14, pp. 1–21. Paris, 1876.
16. BALBIANI, E.-G. *Le phylloxera du chêne et le phylloxera de la vigne*. Imprimerie nationale, Paris, 1884.
17. BALTET, C. *L'art de greffer*. Paris, 1890 edn.
18. BARRAL, J.-A. *La lutte contre le phylloxera*. Paris, 3rd edn, 1883.
19. BAZILLE, G. Des plantes sur lesquelles on pouvrait greffer la vigne. *Bulletin de la Société d'Agriculture de l'Hérault*, Séance 2 Août 1869.
20. BAZILLE, G. *La Vigne Américaine*. 9 Août 1873. Montpellier. *See also* Planchon, Bazille, and Sahut.
21. BERKELEY, M. J. Oidium Tuckerii. *Gardeners' Chronicle and Agricultural Gazette*. London, 1845.
    Berne Convention. *See* Conférence phylloxérique internationale.
22. BERTALL (C.-A. D'ARNOUX). *La vigne*. Paris, 1878; p. 89.
23. BLOCH, R. Abnormal plant growth. *Brookhaven Symposium on Biology* 6, 41–54. 1954.
24. Bordeaux; *Compte-rendu général du Congrès international phylloxérique de Bordeaux 1881*. Feret, Bordeaux; Masson, Paris, 1882; pp. 61, 152, 154, 160, 313.
25. BOUBALS, D. Étude de la distribution et des causes de la résistance du phylloxera radicole chez les Vitacées. *Annales de l'Amélioration des Plantes*, 16, pp. 145–83. Paris, 1966.
26. BOURCART, E. (trans. D. Grant). *Insecticides, fungicides and weedkillers*. Scott, Greenwood & Son, London, 1913; p. 123.

27. BOUTIN, —. Études d'analyses comparatives sur la vigne saine et sur la vigne phylloxérée. *Comptes rendus de l'Academie des Sciences*, xxv, 2e série, p. 17. Paris, 1876.

28. BOYSEN-JENSEN, P. Untersuchungen über die Bildung der Galle. *Mikiola fugi. det Kongelige Danske Videnskabernes Selskab Biologiske Meddelelser*, 18, pp. 1–19. Copenhagen, 1962.

29. BRANAS, J. Viticulture. *VIe Congrès international de la Vigne et du Vin, 1950*. Office International du Vin, 23 Année, No. 237, p. 29. Paris, 1951.

30. BROWN, A. W. A. *Insect control by chemicals*. J. Wiley & Son, New York; Chapman and Hall, London, 1951; pp. 471–2.

31. BUCHANAN, R. *Culture of the grape*. Cincinnati, 1850 and 1865 (8th edn).

32. BUCKTON, G. B. *Monograph of the British aphids*. Ray Society, London, Vol. IV, p. 64. 1883.

33. *Bulletin de l'association viticole de l'arrondissement de Libourne*, Séance générale, 3 Août 1875; pp. 12–14.

34. *Bulletin de l'Office International du Vin*. Janvier, Février et Avril 1951.

35. BURR, M. *The Insect Legion*. Nisbet, London, 1939; p. 179.

36. BUSH, I. *Der Amerikanische Weinbau und Weinhandel in Wislandy, erster Deutscher Jahresbericht der Staats-Ackerbehörde von Missouri*. Jefferson City, 1872.

37. BUSH, ISADOR, AND SONS. *Illustrated descriptive catalogue of grape vines*. St Louis, 1869.

38. BUSH and MEISSNER. *Bushberg catalogue*. St Louis, Miss., 1883.

39. CHANCRIN, E. *Le vin*. Hachette, Paris, 6th edn, n.d.; p. 126.

40. CHANCRIN, E. *Viticulture moderne*. Hachette, Paris, 1950; pp. 100, 116.

41. Columella. *De re rustica*, I, l.

42. Commission départementale de l'Hérault pour l'étude de la maladie de la vigne. *Phylloxera, expériences faites à Las Sorres. Resultats pratiques de l'application des divers procédés*. Grollier, Montpellier, 1877.

43. Commission du Loiret contre le phylloxera (MM. Maxime de la Rocheterie Grenet and S. Darblay). *Rapport sur les expériences faites en 1877*. Orleans, 1877.

44. Conférence phylloxérique internationale. *Actes de la 3 Octobre–3 Novembre 1881*. Berne, 1881.

45. LE CONTE, J. *American Grape Vines*. Patent Office Report. Washington, 1858.

46. CORNU, M. Études sur la nouvelle maladie de la vigne. *Mémoires présentés . . . à l'Académie des Sciences de l'Institut de France*, XXII, p. 1. Paris, 1874.

47. CORNU, M. Études sur le *Phylloxera vastatrix*. *Mémoires présentés . . . à l'Académie des Sciences de l'Institut de France*, XXVI, No. 1, pp. 43–175. Paris, 1878.

48. CROWE, SIR J. A. Foreign Office, *Miscellaneous Series*, No. 176. H.M.S.O., 1890.

49. CULLER, A. S. *The Grape Culturist*. New York, 1873.

50. CURTIS, J. *Farm Insects*. Blackie and Son, Glasgow, 1860.

51. DARIS, B. T. Phylloxera—as a pest of viticulture in Greece. *Pest Articles and News Summaries*, 16, pp. 447–50. London, 1970.

52. DAVIDSON, W. M., and NOUGARET, R. L. *The Grape Phylloxera in California*. United States Department of Agriculture Bulletin No. 9038, p. 4. 22 April, 1921. Washington D.C.

53. DE GRULLY, P. *Essai historique et politique sur la production et le marché des vins en France*. Giàrd, Paris; Coulet, Montpellier, 1910.

54. DELORME, —. Lettre au President du Comice Agricole d'Aix. 8 Novembre 1867. *Revue Agricole et Forestière de Provence*, 5 Mars 1868.

55. DELPECH, A. Industrie du caoutchouc soufflé. Recherches sur l'intoxication spéciale qui détermine le sulfure de carbone. *Annales d'Hygiène Publique*, XIX, pp. 65–189. Paris, 1863.

56. DENISOVA, T. V. The phenolic complex of vine roots infested by *Phylloxera* as a factor in resistance [in Russian] *Vest. sel'sk. Nauki*, Vol. 10, No. 5, pp. 114–18. Moscow, 1965. Summary in *Review of Applied Entomology*, Vol. 53, p. 626. London, 1965.

DERVIN, G. *See* 190.

57. DESCHARNES, R. *The World of Salvador Dali*. 2nd edn 1968, Macmillan, London.

58. DOWNING, CH. *The Fruit and Fruit Trees of America*. London, 1845; New York, 1849 and 1857.

59. DUCLAUX, E. Recherches sur le *Phylloxera vastatrix. Mémoires présentés . . . à l'Académie des Sciences de l'Institut de France*, XXII (5), p. 1. Paris, 1874.

DUFOUR, D. *See* Ordish and Dufour.

60. DUFOUR, J.-J. *The American Vine Dresser's Guide*. Cincinnati, 1826.

61. DUMAS, S.-B. Sur les moyens de combattre le Phylloxera (sulforcarbonates alcalins). *Comptes rendus de l'Académie des Sciences*, LXXVIII, No. 23, p. 1609. Paris, 1874.

62. ESPITALIER, S. *Ensablement des vignes phylloxérées. Instructions pratiques*. Coulet, Montpellier, 1874.

63. *Evening Standard*. London, 14 June 1897.

64. FALIÈRES, E. L'Industrie du sulfure de carbone. *Journal d'Agriculture pratique*, 29 Mars 1877 et 5 Avril 1877, p. 457.

65. FAO *Production yearbook*. Vols. 12 to 21, 1958–1969. FAO, Rome.

66. FAUCON, L. Submersion des vignes comme moyen de destruction du *Phylloxera. Comptes rendus de l'Académie des Sciences*, p. 784. Paris, 1871.

67. FAUCON, L. *Guérison des Vignes Phylloxérées*. Coulet, Montpellier; Delahaye, Paris, 1874.

68. FAUVERAUX, —. *Histoire de l'abbaye de Cîteaux*. Paris, *c*. 1950.

69. FÉRET, E. *Statistiques générales de la Gironde*. Bordeaux, 1878; p. 480.

70. FITCH, A. *Report on injurious insects*, I, p. 158. New York, 1856.

71. FITZ-JAMES, MARGUERITE DUCHESSE DE. La vigne Américaine. *Revue des Deux Mondes*, Juin 1880, p. 889. Paris.

72. FITZ-JAMES, MARGUERITE DUCHESSE DE. *Le Congrès de Bordeaux*. Dudois, Nîmes, 1881.

73. FITZ-JAMES, MARGUERITE DUCHESSE DE. *La Viticulture Franco-Américaine, 1869–1889*. Masson, Paris, 1889.

74. FLEMING, W. E., and BAKER, F. E. *The Use of Carbon Bisulphide against Japanese Beetle*. United States Department of Agriculture Technical Bulletin No. 478, p. 75. 1935. Washington D.C.

75. FOËX, G. *Cours complet de Viticulture*. Coulet, Montpellier, 2nd edn, 1888.

76. FOËX, G., and VIALLA, L. *Ampelographie américaine*. Paris, 1886.
77. FONSCOLOMBE, B. DE. Description du genre Phylloxera. *Annales de la Société Entomologique de la France*, III, p. 222. Paris, 1834.
78. Foreign Office, London. *Trade Reports*. A. 683.
79. FREE, J. B. *Insect Pollination of Crops*. Academic Press, London, 1970; pp. 193–7.

80. GARREAU, —. Notice sur la déstruction des charançons du blé. *Archives d'Agriculture du nord de la France*, 1854 (2), pp. 195–8.
81. GENEVA: *Le Phylloxera dans le canton de Genève en 1892* (Anon.). Tapponier, Geneva, 1893.
82. GIDE, C. *La crise agricole et les Sociétés de Vinification*. Larose, Paris, 1909.
83. GIRARD, —. Études sur la maladie de la Vigne. *Mémoires présentés . . . à l'Académie de l'Institut de France*, XXV, p. 78. Paris, 1876.
84. GIRONDE: *Rapport et Archives départmentale de la Gironde*, Bordeaux. Dossier 7 M—144 [2]. Récompense nationale instituée par la loi du 22 Juillet 1874. M. Laliman.
85. GRANDI, G. *Dispense di entomologia agraria*. Portici, 1911.
86. GRANT, C. W. *Illustrated Catalogue of Vines*. Iona, New York, 1858.
87. GRASSI, B. *Contributo alla conoscenza delle Filloserine*. Rome, 1912; p. 391.
88. GUYOT, J. *Études sur les vignobles de France*. Vol. III. Paris, 1868(?).

89. HALLGARTEN, S. F. *Rhineland Wineland*. Arlington Books, London, 1965; pp. 77–84.
90. HALLGARTEN, S. F. *German Wine Law*. Harper Trade Journals, London, 1969; pp. 35–40.
91. HEUTZE, G. *Revue Horticole*. Paris, 1852; pp. 168–70.
92. HORNE, A. *The Fall of Paris*. Macmillan, London, 1965.
93. HUSMANN, G. *The Cultivation of the Native Grape*. New York, 1866.
94. HUSMANN, G. *American Grape growing and Wine making*. New York, 1896.

95. HUSMANN, G. C., SNYDER, E., and HUSMANN, F. L. *Testing* Vinifera *grape varieties on* Phylloxera *resistant rootstocks in California*. United States Department of Agriculture Technical Bulletin No. 697. 1939. Washington, D.C.

96. HYAMS, E. *Dionysus*. Thames and Hudson, London, 1965; p. 228.

97. HYAMS, E. *The Wine Country of France*. Lippincott, New York, 1960.

98. *L'Illustration*, Paris. *a*. Vol. XLIX, 1867; Vol. XL, 1868. *b*. Le Phylloxéra en Champagne (C. Crespeaux). 18 Août, 1894; p. 135.

99. IRVINE, W. *Apes, Angels and Victorians*. McGraw-Hill, New York; Weidenfeld & Nicolson, London, 1955.

100. *Journal d'Agriculture Pratique*, Paris. *a*. 1867, Vol. 4, p. 693; *b*. 1868, Vol. 1, p. 856.

101. *Le Journal Illustré*, 16 Mai 1875, p. 154. Paris.

102. *Journal Officiel*, 30 Octobre 1875, p. 3, col. e. Paris.

103. JUROVIC, MILOS. Fyloxéra unicova. *Agrikultura Sbornik Pol'nohos-podarskeho Muzea v Nitre*. Nitre (Czechoslovakia).

104. KELPERIS, I. Experiments on the phyto pharmaceutical control of *Phylloxera*. *Bulletin de l'Institut de la Vigne*, I, pp. 3–21. Athens, 1967.

105. Kew Royal Botanic Gardens. *Bulletin of Miscellaneous Information*, Vol. cxxxix, No. 37, 1890. Maqui berries for colouring wine.

106. Kew Royal Botanic Gardens. *Miscellaneous Reports*, 1872–1900.
   a. Letters: notes Nos. 6, 12, 13, 56, 72. J. D. Hooker, Viscount Enfield, Sir Chas. Murray;
   b. Miscellaneous Reports 1–52. Papers relating to *P. vastatrix*;
   c. Davis, No. 124;
   d. Apolis, Nos. 157–61.

107. LAFITTE, PROSPER DE. *Quatre ans de luttes pour nos vignes et nos vins de France*. Masson, Paris; Féret et Fils, Bordeaux, 1883. *a*. pp. 231–9; *b*. p. 345; *c*. p. 477.

108. LALIMAN, L. *Coup d'œil agricole et social, réformes viticoles, cépages indigènes de l'Amérique.* 1860.

109. LALIMAN, L. Sur l'immunité de certains cépages Américains dérivant du *Vitis æstivalis*. *Messager du Midi*, 13 Novembre 1869; 27 Juin 1870. Paris.

110. LAURENT, R. Les vignerons de la côte d'Or au XIXe siècle. *Société des Belles Lettres*, Paris, 1958. Vol. I, pp. 325–62; Table LII, p. 172. Vol. II, p. 55.

111. LEROY-BEAULIEU, P.-P. Le phylloxéra, ses progrès et la legislation. *L'Economiste français*, 4 Août 1879; p. 154.

112. LE ROY LADURIE. *Histoire du Languedoc.* 'Que sais-je' No. 958. Presse Universitaire, Paris, 1962.

113. LICHTENSTEIN, J. *L'Année scientifique et industrielle.* 18e année, 1874; p. 364. Paris.

114. LICHTENSTEIN, J. Notes pour servir à l'histoire des insectes du genre *Phylloxera*. *Annales Agronomiques*, II, pp. 127–42. Paris, 1876; *Annales de la Société Royale d'Entomologie de Belgique*, XIX, pp. 1–16. Brussels, 1877.

115. LICHTENSTEIN, J. *Histoire du phylloxera.* Coulet, Montpellier; Baillière, Paris, 1878.

116. LOUDON, J. C. *An Encyclopaedia of Plants.* London, 1829.

117. M'MAHON, B. *The American Gardener's Calendar.* Philadelphia, 1806.

118. MARCILLA ARRAZOLA, J. *Tratado práctico de viticultura.* Vol. I, p. 52. Madrid, 1954.

119. MARÈS, H. (editor). *Phylloxera, résultats pratiques de l'application des divers procédés présentés aux concours des prix de 20,000 et de 300,000 fr.* Grollier, Montpellier, 1877.

120. MARION, A.-F. *Action Insecticide des Sables.* 1878.

121. MARSHALL, A. (editor). *Salome Dear NOT in the Fridge.* Allen and Unwin, London, 1968; p. 38.

122. MARTIN, C.-F. *Les tables de Martin.* 2nd edn, Paris, 1820. Chapter I, pp. 6–112; Chapter II, p. 19.

123. MAURIAC, F. *Destins.* Arthème Fayard et Cie, Paris, 1935; p. 126.

124. MAYET, V. *Les Insectes de la Vigne.* Coulet, Montpellier; Masson, Paris, 1890; pp. 48, 67.

125. MAYNE, MARGOT. French police battle with wine growers. *The Guardian*, London, 5 February 1971.

126. MILES, P. W. Insect secretions in Plants. *Annual Review of Phytopathology*, Vol. 6, pp. 137–64. Palo Alto, California, 1968.

127. Ministère du Travail. *Annuaire statistique*. Imprimerie nationale, Paris, 1860–1911.

128. MONRO, H. A. V. *Manual of Fumigation for Insect Control.* FAO Rome, 1961; p. 99.

129. MONTILLOT, L. *Les Insectes Nuisibles*. Ballière, Paris, 1891.

130. MORGAN, C. *The Voyage*. Macmillan, London, 1940.

131. MORROW, D. W. JNR. The American impressions of a French botanist, 1873. *Agricultural History*, Vol. 34, pp. 71–6. Washington D.C., 1960.

132. MOUILLEFERT, —. *Le Phylloxera, Comité de Cognac; résumé des résultats obtenus de 1874–1877*. Maison Rustique, Paris, 1877.

133. MUROZ DEL CASTILLO. La plaga filoxérica. *El insecto y la vid.* Logroño, 1878.

134. NACHEV, P. The role of mites in accelerating the destruction of plantations of grape-vines on their own roots. *Review of Applied Entomology*, No. 2426. London, 1969.

135. ORDISH, G. A hundred years of lime sulphur. *Agriculture*, pp. 111–15. London, 1951.

136. ORDISH, G., and DUFOUR, D. Economic bases for protection against plant diseases. *Annual Review of Phytopathology*, Vol. 7, pp. 31–50. Palo Alto, California, 1969.

137. DE PENANRUN, —. Sur la nouvelle maladie de la vigne. *Bulletin de la Société d'Agriculture de Vaucluse*, 7 Juillet 1868, pp. 258–62.

138. PEROLD, A. I. *A treatise on Viticulture*. Macmillan, London, 1927; p. 383.

139. PEROV, N. N., and MIZONOVA, M. N. The effect of fumigants on grape plants and soil microflora. *Doklady Vsesoyoznoi Akademii Sel'sko-khozyaistvennykh Nauk*, IV, pp. 20–3. Moscow, 1968.

140. PLANCHON, J.-E. Le Phylloxéra en Europe et en Amérique. *Revue des Deux Mondes*, Paris. XLIV année, 1 Février 1874, pp. 544–65; 15 Février 1874, pp. 914–43.

141. PLANCHON, J.-E. *Les vignes amèricaines, Leurs culture, leur résistance au Phylloxéra, leur avenir en Europe*. Coulet, Montpellier, 1875.

142. PLANCHON, J.-E. La question du phylloxéra en 1876. *Revue des Deux Mondes*, Paris. XVLII année, 15 Janvier 1877, pp. 241–7. (Often mistakenly quoted as 1887.)

143. PLANCHON, J.-E. Viticulture; sur l'extension actuelle du *phylloxera* en Europe. *Comptes rendus de l'Académie des Sciences*, LXXXIV, p. 1007. Paris, 1880.

144. PLANCHON, J.-E., BAZILLE, G., and SAHUT, F. Rapport à la Société d'Agriculture de l'Hérault sur la nouvelle maladie de la vigne. *Messager du Midi*, 22 Juillet 1868

145. PLANCHON, J.-E., BAZILLE, G., and SAHUT, F. Sur une maladie de la vigne actuellement régnante en Provence. *Comptes rendus de l'Académie des Sciences*, LXXII, p. 639. Paris, 1872.

146. PLANCHON, J.-E., and LICHTENSTEIN, J. Première invasion du phylloxera dans l'Hérault, à Lunel-Viel. *Messager du Midi*, 7 Juillet 1870. Paris.

147. PLINY (G. P. SECUNDUS). *Natural History*. Trans. H. Rackham. Heineman, London, 1960.

148. RAFINESQUE, C. S. *The American manual of grape vines and the art of making wine*. Philadelphia, 1830.

149. RAVAZ, L. *Notes sur recherches sur l'influence spécifique réciproque du sujet et du greffon chez la vigne*. Villars, Paris, 1910.

150. *Regulamento para a venda de sulfureto de carbone*. Lisbon, 1894. Decrees of 30 September 1892 and 16 June 1894.

151. RENDU, V. *De la maladie de la vigne dans le Midi de la France et la nord d'Italie*. Imprimerie nationale, Paris, 1853; p. 95.

152. *Rheinische Volkszeitung*, 22 March 1927.

153. RILEY, C. V. Grape vine leaf Gall. *American Entomologist*, I, p. 248. Saint Louis (Miss.), 1868.

154. RILEY, C. V. *Annual report on the noxious, beneficial and other insects of the state of Missouri*. Began-Carter, Jefferson City. 1869–1874.

155. RILEY, C. V. (editor). *Insect Life*, Vols I–VII, 1888–95. United States Department of Agriculture, Washington D.C.

156. RILEY, C. V. Les espèces américaines du genre *Phylloxera*. *Comptes rendus de l'Académie des Sciences*, LXXIX, pp. 1384–8. Paris, 1871.

157. ROSS, CAPT. SIR J. *Narrative of a second Voyage in search of a North-West Passage*. London, 1835; p. 205.

158. SAVIN, G., *et al.* Rezultate obtinute în combaterea filoxeri galicole la vitele portaltoi. *Anal. Sect. Prot. Pl. Institutului central de Cercetǎri Agricole*, I, pp. 163–74. Bucharest, 1963; *Review of Applied Entomology*, Vol. 55, No. 364. London, 1951.

159. SCHAELLER, G. Biochemical analysis of aphid saliva and its significance for gall formation. *Zoologische Jahrbücher Abteilung für Anatomie und Physiologie der Tiere*, Vol. 74, pp. 54–87. Jena, 1968.

160. SEMPÉ, H. *Régime économique du vin*. Gounouilhou, Bordeaux, 1898; Tables I and XVI, pp. 4, 16–17, 89, 234, 275, 284.

161. SHIMER, H. *Proceedings of the Academy of Natural Sciences of Philadelphia*, January 1867. Philadelphia.

162. SICHEL, A. *The Penguin Book of Wines*. Penguin Books, Harmondsworth 1968; p. 77.

163. SIGNORET, V. Le *Rhizaphis vastatrix* Planchon doit être placé dans le genre *Phylloxera*, Boyer de Fonscolombe. *Bulletin de la Société Entomologique de France*. 12 Août et 23 Septembre 1868; p. 569. Paris.

164. SIGNORET, V. *Le Phylloxera vastatrix*, hémiptère-homoptère de la famille des Aphidiens. Cause prétendue de la maladie actualle de la vigne. *Annales de la Société Entomologique de France*, 4ème serie, IX. Séance 22 Decembre 1869; pp. 541–96. Paris.

165. Société Vigneronne de Beaune. *Bulletin* 18, p. 8. 1893.

166. STERLING, C. Ontogeny of the phylloxera gall of grape leaf. *American Journal of Botany*, Vol. 39, pp. 6–15. 1952. Bloomington, Indiana.

167. STEVENSON, A. B. Seasonal development of the foliage infestations of grape in Ontario by *P. vitifoliae* (Fitch). *The Canadian Entomologist*, Vol. 98, pp. 1299–1305. Ottawa, 1966.

168. STEVENSON, A. B. Soil treatment with insecticides to control the root form of the grape phylloxera. *Journal of Economic Entomology*, Vol. 61, pp. 1168–71. 1968.

169. STEVENSON, R. L. *Travels with a Donkey in the Cevennes*. Chatto and Windus, London, 1916; pp. 170 and 175.

170. STRONG, W. C. *Culture of the Grape*. Boston (Mass.), 1866.

171. TARGIONI-TOZZETTI, G. *Alimurgia*, 1766.

172. *Le Temps*, Paris. *a.* 1 Juillet 1874; *b.* 8 Août 1876.

173. THÉNARD, P., BARON. Essai de traitment de la vigne par le sulfure de carbone. *Bulletin de la Société des Agriculteurs de France*, 1870, p. 391. Paris.

174. *The Times*, London. *a.* 15 August 1872, p. 5d
b. 2 December 1872, p. 12d
c. 24 January 1873, p. 52
d. 10 July 1873, p. 72
e. 15 November 1873, p. 4c
f. 24 November 1874, p. 10c
g. 11 October 1881, p. 9e
h. 13 October 1881, p. 4c
i. 24 November 1881, p. 4f
j. 30 May 1882, p. 3c
k. 30 July 1882, p. 5d
l. 5 August 1890, p. 8e
m. 8 August 1890, p. 8e
n. 21 August 1891, p. 15d
o. 23 March 1892, p. 4f
p. 4 October 1898
q. 25 April 1908, p. 6e, f

175. *Le Tocsin*, 12 Mai 1907. Argelliers.

176. *Trade and Navigation of the United Kingdom, Annual Statement of.* H.M.S.O., London, 1860–71.

177. VANNUCCINI, V. Étude des terres où la vigne indigène résiste au *Phylloxera*. *Messager Agricole*, 10 Septembre 1881. Thonon.

178. VIALA, P. *Les Maladies de la Vigne*. Coulet, Montpellier; Delahaye et Lecrosnier, Paris, 2nd edn, 1887.

179. VIALA, P. *Une mission viticole en Amérique*. Coulet, Montpellier; Masson, Paris, 1889.

180. VIALA, P., and PACOTTET, P. Notes et recherches sur l'influence du greffage. *Revue de Viticulture*, Paris, 1912.

181. VIALLA, L. *Le Phylloxera et la nouvelle Maladie de la Vigne*. C. Coulet, Montpellier; Maison Rustique, Paris, 1869.

182. VIARD, E. *Traité général des vins et de leurs falsifications*. Paris, 1884; pp. 312, 313.

183. VIDAL, —. *Monographie de la commune d'Aimargues*. Paris, 1904.

184. *Le Vigneron Champenois*. 7 Mars 1877.

185. WARICK, R. P., and HILLEBRANT, A. C. Free amino-acid contents of stem and phylloxera gall tissue cultures of grape. *Plant Physiology*, Vol. 41, pp. 573–8. Lancaster, Pa., 1966.

186. WESTWOOD, J. O. New Vine Diseases. *Gardeners' Chronicle and Agricultural Gazette*, 30 January 1869, p. 109. London.

187. WINKLER, M. A. J. Etats-Unis d'Amérique. Rapport national. *VIe Congrès international de la Vigne et du Vin 1950*. Office International du Vin, 23 Année, No. 237, p. 50. Paris, 1951.

188. ZEIDLER, O. Verbindungen von Chloral mit Bromund Chlorobenzol. *Berichte der Deutschen Chemischen Gesellschaft*, Vol. 7, pp. 1180–1. Berlin, 1874.

189. ZOTOV, V. V., *et al.* Physiology of grapes; resistance to Phylloxera. *Sel'skokhozyaistvennaya Biologia*, vol. 1, pp. 410–420. Moscow; also *Chemical Abstracts*, p. 551n. Eaton, Pa. 1966.

(ADDENDUM)

190. DERVIN, G. *Six semaines en pays phylloxerés, étude sur la defense et le reconstituion des vignobles français*. Reims, 1896.

Index